Galileo in Pittsburgh

Galileo in Pittsburgh

Clark Glymour

Harvard University Press
Cambridge, Massachusetts
London, England
2010

Library of Congress Cataloging-in-Publication Data

Glymour, Clark N.
Galileo in Pittsburgh / Clark Glymour.
p. cm.
Includes bibliographical references and index.
ISBN 978-0-674-05103-4 (alk. paper)
1. Science—Study and teaching—United States.
2. Science—Methodology.   3. Education—United States.
I. Title.
QA13.G59 2010
500—dc22   2009030650

For A. L. K.

Contents

III. Science?

Preface

All of the stories in this book are as true as I can make them.

I have had the unusual good luck to take a very minor part in a wide range of scientific investigations, from wildfire prediction to planetary science to psychology to genomics. Each produced its own ironies. I owe much of that opportunity to Ken Ford, director of the Institute for Human and Machine Cognition in Pensacola, Florida, and formerly associate director at NASA Ames Research Center. For many years the institute provided me with a unique combination of research projects.

For many years, Alison Gopnik has been both a sounding board and an unending source of new ideas. Via my student Trevor Thompkins, Claire Ernhart kindly provided a copy of her complaint, with Sandra Scarr, against Herbert Needleman, and Professor Needleman kindly provided me a copy of the report of the hearing board in his case. Emily Scheines provided me with valuable information about Teach for America. As some of the following chapters will suggest, I owe a considerable debt to my colleague Richard Scheines, and a comparable debt to Choh Man Teng, my colleague at the Florida Institute for Human and Machine Cognition. Over many years John Bruer has supported and encouraged the fundamental work that made possible my contributions to several of the topics in this book. Bruce Glymour gave me comments, more penetrating than mine, on education and evolution. I am also indebted to Ted Roush at NASA Ames, who is as honest and brave a scientist as I have ever met. I thank Alison Kost and Esther Kost for permitting me to use and adapt conversations with their late father and husband, respectively, Bert Kost. Madelyn Glymour removed commas, but not enough. Patricia Rich edited the entire manuscript and checked facts. Lindsay Waters, at Harvard University Press, was the editor of the volume of the Minnesota Studies in the Philosophy of Science that debated my very first book, and for a nice circle, of this. I owe him a special thanks for championing a book some readers thought too personal or too candid to see daylight at an academic press. His assistant, Phoebe Kosman, was phenomenal. Lynne McFall took the entire manuscript apart, reorganized it, edited it, told me what was good and what was clumsy. Since she writes so brilliantly, I cannot repay the debt. Peg Anderson, who edited the copy, taught me a lot about how to write.

The book begins with "What This Book Is About," but the reader will soon discover that there is no *thing* the book is about. Consequently, there are a lot of disappointed candidates for the title, beginning with *Light in Shallow Waters* (too poetic), followed by *Paradigms and Bandwagons* (too boring), followed by *Confessions of a Scientific Dilettante* (too true), and, finally, the present title, demanded by my daughter and recommended independently and spontaneously by Lynne McFall. I have neglected the usual academic citations, and only quotations are explicitly referenced. Readings relevant to each chapter are suggested at the end of the book. The curious or dubious reader can check my claims and arguments with them and a little help from Google.

Parts of this book were originally drafted as a project for the National Science Foundation on the challenges that computerized aids pose for maintaining scientific integrity. I am sure the work did not turn out quite as the foundation expected, but I am grateful to the NSF and to the American public for making possible its writing. Neither the foundation nor anyone else except me is responsible for the many opinions that follow.

Galileo in Pittsburgh

What This Book Is About

One day a few years ago a former president of the University of Pittsburgh, a completely serious man who, with his politician's eyebrows, had risen all the way to the rank of major in the army before becoming an academic administrator, decided to visit every department of his institution, in reverse alphabetical order. Philosophy followed physics. The professors of philosophy assembled in beards and mismatched coats and ties to hear with astonishment the president's first challenge, which I paraphrase: "I have just come over from Physics. Those people think there are little, teensy-weensy objects too small to see. You philosophers need to set them straight." He wasn't kidding.

Thirty-some years before, the president of Duquesne University, a Catholic institution in Pittsburgh, published a book, *Euclid or Einstein,* claiming to prove that non-Euclidean geometry, the mathematical foundation of Einstein's theory of relativity, is impossible.[1]

1. Standard hyperbolic geometry, the first non-Euclidean geometry, endorses four of the five usual postulates of Euclidean geometry but denies the fifth postulate, one version of which says the following: given three noncollinear points in a plane, there exists a *unique* line in the plane through one of the points parallel to the line determined by the other two points. In 1931, Father J. J. Callahan proved in his *Euclid or Einstein* a result known several hundred years previous to his writing of it: the first four of Euclid's axioms imply that there *exists* a line through the first point parallel to the line through the other two. After pages of mishmash, he

The president of Duquesne was beating a drum on the tail end of an antirelativity bandwagon. That parade, joined by various astronomers, physicists, and philosophers, started almost as soon as Einstein published. What marked it as a bandwagon was misunderstanding— misunderstanding of what Einstein's theory said, misunderstanding of the assumptions of experimental methods, misunderstanding of the scope of theory, or, as in Father J. J. Callahan's case, misunderstanding of the basic mathematics. Pittsburgh's president wasn't on a bandwagon—he was just silly.

In one of the most influential books ever penned about science, Thomas Kuhn wrote about scientific "paradigms," meaning, roughly, scientific accomplishments whose doctrines and methods came to serve as models for other efforts. In the seventeenth century, Isaac Newton's *Principia* provided a paradigm for celestial mechanics and later for investigations of electrical forces. A century or so later, Antoine Lavoisier's experiments to establish the role of oxygen in combustion provided a paradigm for the experimental method in chemistry. Less famously, George Udny Yule's study of nineteenth-century pauperism using the statistical method of regression formed a paradigm followed today in virtually every science that uses data obtained without experimental controls.

A bandwagon is the evil alter ego of a paradigm, a widespread misapplication or misunderstanding of a theory or method or doctrine— or just a popular mistake. Even the best of theories can serve as the source of both paradigm and bandwagon. Newton's did. While the Marquis de Laplace developed Newton's framework rigorously, others, notably the eighteenth-century affinity chemists, took Newton's writings as a warrant for talking of ill-defined and ill-measured affinities between chemical substances. Chemical affinities were a bandwagon. The statistical method of regression, brought to popular attention by

---

then concluded that he had demonstrated that it is inconsistent to accept Euclid's first four postulates but to deny the fifth, and since hyperbolic geometry does exactly that, he claimed to have proved hyperbolic geometry inconsistent. Hyperbolic geometry does entail the existence of the required parallel line, but in a plane it allows infinitely many lines parallel to a given line.

economists' studies of social phenomena, is a combination of paradigm and bandwagon, used in causal investigations from economics to epidemiology, sometimes appropriately but often not, in ways I will describe.

We want to be within a paradigm not on a bandwagon, but often the hard thing in science is to separate the two. Modern science has made that decision harder, much harder, for many reasons. Twentieth- and twenty-first-century science is more esoteric than its predecessors—atomism is comprehensible, but string theory is explained only with metaphors. Modern scientific doctrines and their evidence can come from a vast consortium of diverse expertise, not much of which can be mastered by anyone. Is the study of human causes of global warming, and of global warming itself, a new paradigm or just a bandwagon? For reasons the reader will discover, my vote is for a paradigm, but few of us are in the position of having a judgment based on more than a will to trust some people but not others, and even each expert knows firsthand comparatively little of what must be trusted to conclude that we are heating up the planet. The prediction of planetary climate is a novel methodology, involving extrapolation of chemical experiment, observation (without experiment) of previously unthinkable quantities of data, computer analysis of the data, and extensive computer simulation and testing. We have not yet quite absorbed the complexity of our own science or what our uncertainty about it means for practical policy. That is part of what this book is about, however obliquely.

Contemporary science is democratic and overpopulated. Journals abound and new ones are created every day, which means there are professional eddies unnoticed by those not trapped in them, and the contradictions between work in one journal and work in another may never be noticed. Contemporary science is computational and automated. A statistical analysis can be done nowadays by pushing a button, and the button pusher need know nothing of the underlying computations or the conditions, or even the meaning, of the results the computer produces. A biological study or a climate study can depend on a computer program whose code is never published and, if it were published, it would be so ill documented as to be unfathomable.

To identify fundamental errors in claims about cold fusion seems easy in comparison to assessing doctrines assembled from hundreds, even thousands, of studies using diverse methods published in a variety of journals or studies done at great expense with one huge sample and unlikely ever to be repeated. Science, like every other collective human enterprise, has always depended on trust, but nowadays our trust must be spread more widely, with less opportunity for independent testing of claims.

Much of this book is a collection of personal anecdotes and reflections on some paradigms and bandwagons in contemporary science and education, for the most part from scientific enterprises I have had a more or less direct hand in. It is chiefly about *applied* science, the kind of science that makes a difference to daily life, not the profound sciences that the university presidents misunderstood. It is about attempts to shed light in shallow waters. And it is about how the factors that make our science different from the sciences of previous centuries may, and sometimes should, erode our trust. Our educational system provides the next generation of our scientists—at least those we do not import. Some people even aim to study education scientifically. The corridor between science and education lets one into the other, in both directions, and so does this book. I admit there is one chapter about education that has nothing to do with science. But it is a true, funny story, a metaphor for the absurdities that follow, and so I begin with it.

I
Education

1

Socialism in One (Indian) Nation

All parents anxious for their children fuss about American public schools if private schools are unavailable or out of financial reach. Everyone has a story, or more than one. My parents, for example, had a story about how their daughter, my youngest sister, became a Mormon convert when they lived in Utah. The local high school had a simple strategy. Sophomores had an elective: they could take either a course in Mormon theology or a course in quantum field theory. True story, approximately, but here is my story about one of my children. Let it be a lesson.

About thirty years ago I resigned from my job at Princeton, where my (former) wife was very unhappy, to take a position at the University of Oklahoma. We moved to Chickasha, about forty miles from Norman, the university town. My wife was a lot happier in Chickasha than in Princeton. I was not. There were three problems: the Language problem, the University problem, and the Chickasha problem, all of which could, I suppose, be combined into one big Cultural problem. The Language problem was that I did not speak or understand Oklahoman, and in small, everyday matters I needed to find a translator, some bilingual talent who spoke both Oklahoman and American English. (I admit I have an ungifted ear; once a distinguished, garrulous visitor from Oxford, Gwil Owen, an expert on the philosophy of ancient Greece, finding no one more important available in my department, held forth in my office for an hour while I listened desperately, not understanding a single sentence, replying with "umm" whenever his voice rose and I guessed he was asking a question.)

A typical Oklahoma example: when I was buying a vase at the Chickasha Wal-Mart (where else?), the woman behind me in the checkout line addressed me as follows: *Bea pup seabyya.* Puzzled, I asked her to repeat herself ("Come again?") and she did: *Bea pup seabyya.* Several people nodded in emphatic agreement, some of them repeating *Bea pup seabyya,* which only made me more worried that I had missed some important communication. *Bea pup seabyya?* Was she saying the vase would leak? Admiring my taste? Were my trousers unzipped? Finally, sensing my difficulty, a bilingual approached and translated for me: *She's saying you had better put it up on the seat beside you.* You know, not on the floor or the bed of your pickup. Jot it down in your phrase book if you ever travel that way: *Bea pup seabyya = Better put it up on the seat beside you.*

The University problem was that there was no university, really, just buildings and unemployed people with salaries and a few unfortunates who were serious scholars, most of them desperately trying to find jobs elsewhere. The university library was an enormous building with three stories. Soon after arriving, I went to find a copy of one of the standard classics of philosophy, David Hume's *Treatise of Human Nature.* Not finding it in the card catalog, I inquired at the reference desk, which yielded the information that *that sort of book* was available only through interlibrary loan. This provoked a certain curiosity. The spacious first floor of the library was taken up with the usual for the time: a reference section with encyclopedias and such, a reading area, a card catalog, a few shelves with actual books, staff offices. I took the elevator to the top floor, only to find there was nothing there. Completely empty. I took the elevator down to the second floor. It was not empty, just empty of books. The second floor was filled with endless cartons of out-of-date emergency rations issued by civil defense in case of nuclear war. When the bomb fell, Norman was ready to eat. (A year later, when I did own a pickup, I and some friends carried out tons of cartons of those rations to feed to our pigs. I don't know if they were replaced by books or by more up-to-date rations.)

There were some professional disappointments to Oklahoma university life. As soon as all of the faculty members had taken their seats at the first department meeting I attended, one of my new colleagues rose to announce that everyone he had ever met from Princeton was a pinhead. He did not give an enumeration of his other Princeton acquaintances. That colleague—his name was Bill—provided my finest moment at Oklahoma after another meeting later that year, during which he raged at me for failing a graduate student's qualifying essay in ethics. (She had written about John Rawls, the late, justly celebrated author of *A Theory of Justice,* chiefly complaining at length that Rawls's theoretical device, the *veil* of ignorance, was patriarchal and antifeminist.) When the chairman tried to quiet Bill, he raged at the chairman for assigning someone utterly unqualified, namely me, to read the examination. The finest moment came the next day. As I was unlocking my office, Bill hailed me from down the corridor and said he wished to speak with me. I let him into my office and took a seat— he stood—and, with perfect sincerity, he apologized: *I want you to understand, Clark, I wasn't angry at you—I was angry at the chairman for letting an incompetent grade the exams.* I thanked him for the apology. How good can it get?

Woody Allen has a two-liner: *Consciousness is a wonderful thing. I wonder what they do in New Jersey?* Some places are in fact more alert than others. It has nothing to do with size. Missoula is as awake as New York, just different, and in its heyday Butte, Montana, where I grew up, was positively all nerves. Chickasha was catatonic. Cherokee and Choctaw Indians, some of them my wife's distant kin, in from the nearby reservations, provided the only variation from a kind of double-knit Baptist placidity, and they were given a wide berth by the nonnatives.

Oklahoma innumeracy reached its zenith—or nadir, however that works—in Chickasha. The chairman of the high school English Department asked my wife to explain to the English teachers how to average grades; the division part really got them. Restaurants featured pizza exclusively, as though it had been invented there, and music was available on the radio if you liked country cryin' and dyin', as, I admit, I sometimes do.

So when my son came home from school one afternoon to find me alone in the tasteless, graceless, three-bedroom tract house we had rented in residential Chickasha, I was not happy. From the mood radiations twelve-year-olds give off I could tell he was not happy either, which was surprising, since he was a cheerful boy. I asked him if something had gone wrong at school—he was in his last year at Chickasha Junior High School—and he said something had indeed gone wrong. He had been made to stand in the corner, facing the wall, for the entire hour of his art class because he had forgotten to bring a pencil. I found this unlikely, but I don't believe in dithering, so I called the school and asked to speak with the art teacher, and did. An educated man for Chickasha, he spoke American English. The conversation went something like this:

> Hello, I'm Clark Glymour. You have my son in art class. He came home today and said you made him stand in a corner for an hour, and I'd like to know more about it.
> That's not true, Mr. Glymour, I didn't make him stand in the corner.
> I'm glad to hear that. I understand that children often don't give entirely accurate reports. Why don't you tell me what did happen?
> Well, Mr. Glymour, I gave him his choice. He could have stood in the corner or against the wall.

I could not fault the rigorous logic, but as in the essays of many contemporary philosophers, the rigor seemed to miss the point.

> I think there is a moral issue here we need to talk about. I'd like to come down and speak with you.
> OK. I'm in the art room on the second floor.

Leaving my son at home with a Coke and no smile, I drove to the junior high school, preparing myself along the way. My sole experience with art teachers inclined me to think they are a tough breed given to extremes in crowd control. In Butte High School art had been taught by a round, granddaddy-looking man with long white hair, Pop Weaver, who was alleged to fondle girls in the coat closet of his classroom, but who, whatever the truth of that rumor, undoubtedly had an effective and novel strategy of classroom management. Pretty girls were assigned seats in the front row of desks and jocks were assigned seats in the second row; everyone else could sit where they pleased. Pop carried a big water gun during class, and whenever someone—always a jock—got out of line, Pop splashed him in the face. Once, when an ambitious jock fired back with his own water pistol, Pop upended the fellow's chair, captured the offending weapon from the floor, and announced that he had the only guns in class.

The Chickasha Junior High School art room was a large rectangle with blackboards on one side, windows opposite, shelves for supplies at one end, and a few drafting tables at the other. Three long rows of desks faced the blackboards. The art teacher, whose name was Joe Something, shook my hand in a manly, Oklahoman way. Every finger of each hand was encircled with rings stoned with turquoise; his shirt was Cowboy with mother-of-pearl buttons, and his Levis were secured with an Indian-worked buckle, suitable in size for a boxing-title belt. The man himself was lean, almost scrawny.

I made my pitch: Children should be treated with respect. I, and assuredly he, had attended meetings without a needed pen or pencil—it was an ordinary error, and no adult would be made to stand against the wall throughout a meeting for having made it. Nor should a child.

Well, Mr. Glymour, I had to set an example. If the kids don't bring their pencils, they can't do the art.

I asked him if other students sometimes forgot their pencils, and he said they did, quite often, and he made them stand against the wall or, at their choice, in the corner.

Wouldn't it be better, Mr. Something, if you had each child bring a small box—a cigar box, say—to leave their pencils and other drawing supplies in? That way everyone could do the artwork every day and no one would have to stand against the wall or in the corner.

I can't do that, Mr. Glymour. I've got a hundred and forty-four kids in my classes, total, and there isn't any room for all those boxes.

In proof, he led me around the room, pointing to each spot and testifying that no one of them, not even all of them together, would hold 144 cigar boxes. My entire academic career, I realized, had prepared me for this moment. Only months before I had been a colleague of Thomas Kuhn, whose famous book, *The Structure of Scientific Revolutions*, had introduced the latest academic phrases gone popular— *paradigm, paradigm shift, conceptual framework.* (Now disappearing from popular writing, sad to say, with the paradigm shift to *deconstructed* and *commodified.*) What Mr. Something required was a paradigm shift, a new conceptual framework.

Do you have room for one cigar box?

Evidently puzzled at the question, Joe Something guessed that he did.

Well, then, why not use one cigar box and have each student leave a pencil in it? Let each student take a pencil from the box at the beginning of class and put it back at the end of class. You never have more than thirty or so students at a time, so even if some of the pencils are lost or not returned, there will always be enough for each student to use one.

Joe Something was silent, completely quiet for a full minute, thinking the proposition over carefully. And then:

I can't do that, Mr. Glymour.

I was irritated, the irritation of the Conceptual Revolutionary at the Conceptual Antirevolutionary, and I suppose it showed:

For God's sake, why not?

He answered, and as Dave Barry so often says, I am not making this up:

Because, Mr. Glymour, that's socialism.

2

David Koresh Middle School:
Observations on Education in America

There are a lot of choices available if you want your kid to get a sound education in mathematics and science and you cannot afford a fancy private school. You can send your child to any of several inexpensive countries for a better education than can be expected in America, among them Slovenia and South Africa. That's right. On average, when they leave high school, kids in Slovenia and South Africa—Hungary, too—do better at both science and math than American kids. America graduates a much higher percentage of its youth than does South Africa, which essentially educates only the select (which of course your kid would be); but your kid would also get a better education in almost any European country and in Singapore as well. Of course, they have so many advantages in Slovenia; it isn't really fair to compare them with us.

American elementary and high school education lives up to third-world standards, almost, but we in America spend a lot more money than most countries do to achieve less, about half again what Europeans pay as a percentage of gross national product. Actually, we spend more. The United States maintains, largely at public expense, a vast system of remedial and vocational schools and community colleges, designed to make up for part of what American students missed in high school.

In some of the worst school systems we spend the most. District of Columbia schools claimed (in 2007) to have spent about $13,000 per pupil, but with proposed construction costs the real number was surely much higher; for example, a recent construction budget alone would add about $4,000 per student for the next decade. Pittsburgh, which a local newspaper and an independent research institute say spends about $17,000 per student each year (the school system claims less), has appalling student performance according to a 2003 Rand Corporation report. In Pittsburgh 65 percent of white students and 31 percent of black students were reading at their grade level in 2003; 56 percent of white students and 25 percent of black students were proficient in math at their grade level. Sixty-one percent of the students are black. You do the math.

There is no hiding the correlation of ethnic and economic background with school performance. The more academically ambitious students who are about to graduate take the SAT. SAT scores for Hispanic students average about three quarters as high as those for European American students; the scores of African American students are lower still, even in reading and writing tests in which children raised speaking Spanish might be thought to have a disadvantage. Estimates of nongraduation rates for African American and Hispanic students run to about 50 percent. If GED students are counted, the percentage of those youths who pass age twenty or so without a high school degree or equivalency falls to about 25 percent. About 8 percent of American adolescents are foreign born, but they are 25 percent of teenage school dropouts. Asian students, foreign born or not, do much better on average.

So the questions are *How come?* and *What can be done about it?* Some possible answers:

1. American students are stupider than those in most of the rest of the world. Bring in smarter kids.
2. American parents are a dysfunctional mess. Issue parenting licenses.
3. American public education institutions are an incompetent, dysfunctional mess. Bring on private schools.

4. American teachers are stupider than those in most of the rest of the world. Bring in smarter teachers.

All four answers may be correct. Not all of the solutions are.

1. Let's start with stupid. Stupid is what someone cannot do—talk, write, calculate, solve abstract problems, solve concrete problems, pay attention, reason, understand others' reasoning, remember, respond to questions, manipulate objects, understand others' states of mind and emotions, distinguish what is relevant and what is irrelevant to an issue. Judged by academic performance, on average, American kids are not as smart as those in many other countries on many of these dimensions. The same is true for elementary school kids, although they compare a little better with other nations than do kids in higher grades, suggesting that given time, the American school system makes kids stupider.

Stupidity starts early. The ability or inability to learn in English, to do academic work well, starts early. Don't read to your kid, have no books in the house, don't teach your kid to draw or use a pencil, don't play board games with your kid (Chutes and Ladders is key), let slightly older kids raise your kid, leave the television on night and day, and chances are your child will begin school stupid. Some kids who spend their first five years this way will catch up, but many never will, and for them school will always be part humiliation, part annoyance. At home or in school, actively or by neglect, stupidity is taught.

Since most of what is taught in American public schools is taught in English, stupid is also relative to language. A Los Angeles Unified School District elementary school teacher may face a classroom with children speaking as many as six distinct first languages. Lots and lots of black children come to school speaking something other than English, sometimes called Ebonics, but in any case a language difficult for European English speakers to understand, and very likely the converse is true.

Maybe Chinese and other Asian people are born wired better for mathematics and science than the rest of us, maybe not, but their kids do better in American schools, and we know one reason why: their parents expect them to and help them to, and the kids form the habit.

Rather intrusive studies of Chinese American families show a pattern of organized, systematic parental attention to schoolwork. There is probably nothing essential about being Chinese or Asian. Data from a (rather old) large study of high school students and their families suggests that whether students plan to go to college or not is strongly influenced by parental encouragement—and the influence is independent of student's IQ or family socioeconomic status.

2. A few years ago my eldest daughter formed a friendship with a Vietnamese girl, Anh Ngoc, a "boat person" who, with two younger brothers, had made it from Vietnam to a camp in Hong Kong at the end of the war and from there to the United States, where they had a relative. One hot summer day my daughter gave me something to deliver to Anh. As I drove through her down-at-the-heels neighborhood, I saw boys playing basketball, boys and girls hanging out on sidewalks and porches, doing not much, no books in sight. Anh was on her porch, reading. She invited me into the sweltering, cramped apartment she shared with her brothers. Her desk was in one corner; in another corner a brother dressed only in shorts studied English; in another a brother similarly dressed studied chemistry. (Anh became valedictorian at the most prestigious public high school in Pittsburgh. Carnegie Mellon University held a press conference to announce they were giving her a scholarship. When, back home, Anh opened the envelope and read the enclosed letter, she discovered that the scholarship was for half of the university's tuition. She returned to the dean of admissions with the letter and apologetically notified him that her financial circumstances did not permit her to accept the half scholarship and that she would have to call a press conference to explain why. She received a full scholarship and graduated something cum laude with a double major in engineering and French. Those cunning Asians.)

No doubt sometime, somewhere, other kids read books on their porches in Pittsburgh summers, but in twenty-five years of driving and looking, I have seen only Anh. What I do see are children, black and white, as young as ten or so, raising their parents' children, pushing baby carriages, leading toddlers. We can't license parenthood, and we can't have a nanny in every neglectful household, but of course there is something we could do. We could open our schools twenty-four hours a day, every day of the year, providing warmth and safety and fun and nourishment and civility and instruction for neglected children from infancy through adolescence. "Midnight basketball" was one of William Clinton's initiatives in the 1990s and also one of George H. Bush's forgotten "thousand points of light." Clinton's proposal was lampooned by Newt Gingrich (the man who wrote that our economic future lies in biomedicine, but we shouldn't teach evolution) and Rush Limbaugh (never mind, literally), but midnight basketball is just the rim of the hoop of what we could do. What would open schools cost with some volunteer help—a couple of hundred billion dollars a year perhaps? Most crime is done by young men behaving badly. What does crime cost the United States? One estimate is a trillion dollars each year. Open the schools, knock down crime by half, leave your share of the difference in your pocket—a thousand bucks more or less. And you get a civil society thrown in.

3. American schools are famously uneven for two reasons: most of their financing is local, which means poor people have poor schools, and much of their government is local, which means people get the schools they care about, or don't. Every other industrialized nation (although I'm not so sure "industrialized" applies to the United States any longer) has a national school system, sometimes with private alternatives. American schools are fabulously top- and side-heavy, with counselors, curricular specialists, nurses, psychologists, coaches (who often also teach "social studies"), assistant principals, associate principals, police, whatnot. Ohio, to take a near worst case, reportedly has one administrator for every twenty-one teachers. In all but five states, at least 45 percent of public school employees are nonteachers. Some of the schools are just pass-throughs. A Hartford high school, for example, lets dropouts come for lunch and hang around, no instruction. Maybe that's better than leaving the kids on the streets, but it's not good enough. Teachers' salaries vary from poor (Montana) to quite good (many big cities), even when adjusted for cost of living; they are heavy on benefits. Unfortunately, the quite good salaries have nothing to do with quite good teaching. Salaries increase by years of teaching, by phony degrees, and by administrative role, almost never by teaching effectiveness. Almost all advanced degrees held by teachers are in

education, which means they are academically rock bottom. A phony doctorate and a dozen or so years of experience will produce an income of $100,000 or more—plus benefits—in some urban and suburban high schools. In Pittsburgh a teacher automatically receives a larger salary increase for a doctorate in education than for a doctorate in some substantive subject, say history. Many schools give teachers tenure after two years, which means permanent employment barring major, major missteps. Insanity and paranoia are not firing offenses. A history teacher in a Pittsburgh high school explained to his students that Hurricane Katrina was deliberately caused by the U.S. government using big, mysterious wave machines out in the ocean, presumably to empty New Orleans of "chocolate" people (in New Orleans Mayor Ray Nagin's inapt phrase). When parents complained, the principal told them, in effect, to live with it. (Nagin, the man who left about a thousand buses parked before and during the storm, also blamed the federal government, but not for the storm, which he said came about because God was mad at America—same mechanism the Reverend Jerry Falwell gave for the attacks on 9/11, but God had contrary reasons for his anger in the two cases. Hard Guy to please.)

Public schools are heavily unionized, and teachers' unions have effectively fought against the introduction of salary scales based on rational measures of teaching effectiveness. Parents (and nonparents even more) rail at school taxes but will not endure a strike, and school boards are in the middle. Thirteen states allow teachers to strike, and Pennsylvania leads the nation in the number of strikes—by a lot: 60 percent of the teacher's strikes in the seven years prior to 2008 were in that state.

A teacher influences a classroom, or tries to. A principal influences a school, or tries to. Principals in America take advanced degrees in principaling, whose curriculum and texts, aside from potentially useful information on school law and finances, are largely pop psychology at about the level of the Sunday supplement *Parade,* although not so well written. In many cases principals have a hopeless task, with teachers who cannot be fired and cannot be transferred, students who cannot be disciplined, a rigid curriculum, and an inflexible budget.

Chiefly to encourage high-tax-paying professional families not to flee the school district, many urban high schools are essentially three schools in the same place, one for the black kids who are troublemakers, one for the white kids who score really well on IQ tests, and another for everybody else. An academically serious kid who doesn't test well is doomed. Elementary schools are generally tracked less aggressively, with somewhat different assignments for "gifted" students, occasional grade jumping, and regular removal from the classroom for special projects. About 13 percent of students, disproportionately black and American Indian, are assigned to "special education." Essentially, American public education is by group as much as age, with group membership determined by tests. If classification were entirely by IQ test, or by any of many tests strongly correlated with IQ test scores, the lowest 13 percent would have IQ scores lower than 84—the IQ score of one of my sisters. My sister can add and read, but, like Oklahoma teachers of English, she cannot find the average of two numbers. Children with her abilities need special help.

There are five responses to reforming the structure of American schools: (1) change nothing fundamental but spend more money; (2) let families transfer their children to the public school of their choice; (3) set up charter schools; (4) homeschool children; (5) replace the entire system with private schools. None of them are serious. The American Federation of Teachers and the National Education Association are for more money. Money is the problem in some places, but how money is and is not spent is the problem in many places. Politicians who want to compromise between the teachers' unions and parental demands favor school choice, a solution that doesn't work because there are not enough places in preferred schools and because it leaves behind the children with less concerned parents. Charter schools, which operate with more autonomy from school boards and labor contracts, are of a thousand kinds, with studies showing that they both are and are not better than conventional schools. More than a million children are homeschooled, generally because their parents want to shelter their children from one or another aspect of the school environment—violence, low academic quality, or secularization—and have the time and resources to do so. The majority of homeschooling families are religious nuts. Very right-wing pundits want the entire system privatized, generally on grounds of their fundamentalist belief in the superiority of markets to solve every problem. But two of the functions of public schools are to create a sense of national identity and an equality of opportunity; the first would surely be lost with privatization and the second never gained. And there would be some huge social costs. Do you really want a David Koresh Middle School?

4. Are American teachers the problem with American schools? Public school teachers, by and large, are former college students who majored in education, or in some cases in a science or mathematics supplemented with an education curriculum. On average, the SAT scores of education majors are just a bit above those of football players. (Imagine if football games were scored like this: points scored by team + sum over all players on a team of [time playing × SAT score.]). All kinds of personal qualities and skills we do not know how to measure well undoubtedly contribute to good teaching, and we probably would not want Bill Gates teaching third grade. But it is a reasonable thought that more, smarter teachers might help things. That was the thought behind Teach for America. Begun some years ago as a college student's vision and funded in part by the federal government and in larger part by corporate donors, Teach for America gives a brief—about ten weeks—but intensive training to well-educated college graduates, most without formal teaching certificates, and sends them off to teach for two years at "underperforming" public schools, code for schools where students are doing really terribly. The would-be teachers receive a couple of weeks of student teaching, training in course preparation, and a lot of team boosting. Paid by Teach for America at whatever the prevailing beginning teaching wage is in the school to which they are assigned, many Teach for America instructors could have found better-paying, less stressful, and more permanent employment elsewhere.

Even remunerated, the Teach for America instructors form an idealistic group motivated by a desire to help students in some of the worst schools in America. No wonder, then, that their efforts and their program are despised, reviled, and condemned by professors of education. Bill Johnson, who teaches education at Yale, has claimed that Teach for America disrespects teaching, which means, I think, that the program is premised on the thought that ed schools don't do a very good job of producing a competent cadre of teachers. It *is* premised on that thought, and maybe it's true.

Political scientists, economists, and education professors have published a lot of statistical studies of the effectiveness of Teach for America instructors. The studies have a pattern. Studies by political scientists, economists, and others not directly in the ed business tend to find that Teach for America instructors, on average, get somewhat better results than other teachers of comparable experience and teaching environment. Studies by professors of education find that Teach for America instructors do worse than the graduates of schools of education. The most cited study of the latter kind, by Professor Linda Darling-Hammond of the School of Education at Stanford, is a careful regression analysis of students in elementary grades in the Houston school district. She had a large sample, controlled statistically for about everything she reasonably could control, and concluded that Teach for America instructors, on average, produce less learning improvement than do comparably experienced and situated teachers who have conventional teaching credentials. Most convincingly, she claims that in the program's early years she studied, a high percentage of Teach for America instructors arrived with conventional teaching credentials and achieved better learning outcomes than regular teachers comparably experienced, but in later years fewer than 50 percent of Teach for America instructors were "credentialed" in advance, and their teaching in those years was on average less effective than that of comparable but credentialed regular teachers.

I would have liked to reanalyze Darling-Hammond's data, but although her paper complains about her inability to obtain data from another study, she did not respond to three requests for *her* data. What would I have looked for? This: indications that there are factors not recorded in her data that influence Teach for America instructors differently from regular instructors, factors that produce an association between formal teaching credentials and teaching effectiveness that Darling-Hammond's regression methods mistook as a causal connection.

Sometimes poking around in the facts and listening to people is more informative than statistical algorithms. When Joshua Kaplowitz joined Teach for America in 2000, after the standard training in team building, course preparation, and brief student teaching, he was sent to a fifth-grade class in Emery Elementary in the District of Columbia school system. A new principal had just been assigned there. Kaplowitz immediately recognized that his problem was controlling a few disruptive students; one small misbehavior—throwing a wad of paper—would lead to a cascade of increasing disruptions:

> On a typical day, DeAngelo [a pseudonym, as are the other children's names in this and the next paragraph] would throw a wad of paper in the middle of a lesson. Whether I disciplined him or ignored him, his actions would cause Kanisha to scream like an air-raid siren. In response, Lamond would get up, walk across the room, and try to slap Kanisha. Within one minute, the whole class was lost in a sea of noise and fists. I felt profoundly sorry for the majority of my students, whose education was being hijacked. Their plaintive cries punctuated the din: "Quiet everyone! Mr. Kaplowitz is trying to teach!"[1]

Kaplowitz cared. He tutored students in his class after school hours, but classroom time was increasingly wasted with racial insults and violent disruptions. The school district had a "no corporal punishment" rule, which meant that teachers were not permitted even to physically separate fighting students—they had to call and wait for the school police to arrive. He tried the banal "classroom management" techniques recommended by so many books for teachers. None worked. He sent students to the principal's office—they were promptly sent back. Some fellow teachers with larger classrooms allowed him to send miscreants to their classrooms when students got out of hand. The principal stopped the practice. No disciplinary action of any significance was possible. Just before standardized tests, the principal changed the no-discipline, no-segregation-of-miscreants policy. It was promptly reinstated after the testing sessions were over. But the principal did take an interest in Kaplowitz. Without visiting his classroom,

1. Joshua Kaplowitz, "How I Joined Teach for America—and Got Sued for $20 Million," *City Journal* (2003).

she would pull him out of class to lecture him on what a bad teacher he was. When he turned in grades, the ever attentive principal said they were too low and ordered him to change them. When he refused, she cited him for insubordination. Kaplowitz was moved to second grade, where, to his surprise, student behavior was worse. Fistfights and hair yanking broke out every day during class. Eventually, when a perpetually disruptive student insisted on being allowed to use the bathroom in the middle of class, Kaplowitz released him with his hand on the child's back as he went out the classroom door. Soon after, police arrived, handcuffed Kaplowitz, and took him to the police station, where he was charged with child abuse—he had, after all, *physically* nudged the child out the door, actually touched the kid. Kaplowitz was tried and quickly acquitted, but the student's family filed a twenty-million-dollar—yes, $20,000,000—lawsuit against Kaplowitz and the school district. The district and the teacher's union insurers settled with the family for ninety grand.

Maybe Kaplowitz is just an incompetent whiner? Maybe he exaggerates? So I talked with Emily, who did not know about Kaplowitz. Emily graduated from Cornell in history and joined Teach for America in 2007. Except for the lawsuit, her story is almost exactly Kaplowitz's. After training that emphasized team building, course preparation, visits to a range of schools with orderly classrooms, and brief practice teaching, she was sent to teach in a failing high school in Hartford, Connecticut, whose students are almost exclusively Puerto Rican, Jamaican, or African American. The first day of school she discovered she was teaching world history, U.S. history, civics for students who had failed civics, and two classes preparing students for the SAT. So much for *her* preparation. The superintendent had arranged for the hiring and assignment to schools of the Teach for America instructors, and the principal was not eager to have them. Nor were many of the teachers. Emily reports that many of her fellow Teach for America instructors do their best to keep secret the nature of their employment. She reports the same kinds of disciplinary problems as did Kaplowitz and the same lack of support from principals and assistant principals. Students wander the halls, absenteeism is astronomical, the ID system is broken, and dropouts come in for lunch and hang about but attend no classes. Like Kaplowitz, Emily functions as a social worker as well as a teacher. When a serious student stopped coming, Emily found out why: the girl was staying at home to care for a preschool-age sibling. Emily arranged for a babysitter. She phones a habitually late student every morning to wake him for school.

I talked with Emily over a Christmas holiday when she was refreshed, but she reported being exhausted. No surprise, with teaching, preparation for teaching, and trying to make sure that outside of class her students get some of the social and educational support they need. She hasn't been sued—yet. She reports that a friend working in Teach for America has been. The suit is for failing to provide instruction for an autistic child fully in accord with agreed procedures. The school administrators had never notified Emily's friend of the required procedures, so she was winging it.

Which brings us back to Professor Darling-Hammond. Crummy principals make for crummy schools, no matter what else happens. Principals who join Bill Johnson in their dislike for Teach for America can make things miserable for the young teachers, and, judging by the two examples, do so. So we have one straightforward variable that could explain the difference in learning outcomes Darling-Hammond found in Houston: administrators behave differently toward Teach for America novices than toward new teachers with conventional credentials.

That would not explain Darling-Hammond's finding that in Houston the average learning outcomes for Teach for America instructors dropped as the proportion of them with conventional teaching credentials declined. Surely that's evidence of cause and effect! Indeed it is, but what is the mechanism? Is it that, other things being equal, teachers who have passed through the conventional, and conventionally dismal, curriculum of a school of education and spent a semester in practice teaching are better teachers than Teach for America instructors who have not? Or is it something else?

My bet is on something else. Recall Emily's exhaustion. At a rough guess, she is working about sixty hours a week. No reason to think that's unusual for novice, socially committed teachers, whether in

Teach for America or not. But in the Houston school district Darling-Hammond studied, there was a difference. Teachers without credentials had to enroll in a local Catholic college to obtain a teaching certificate during their first two years. Houston Teach for America instructors without credentials are doing Emily's job, plus they are going to school. Suppose (and I am only supposing—I do not know) that takes an additional ten or fifteen hours a week during the school year. Now you have young teachers working seventy to seventy-five hours a week. Talk about exhausted! So here is a hypothesis: as the proportion of already credentialed Teach for America instructors in Houston declined, the proportion of them working killing hours increased. Exhausted teachers are bound to be worse teachers.

So what are we to do about American public schools? This: one payer, one set of standards, all-day and all-night schools for every age, a rational system of accountability. But we probably won't because, as the art teacher said, that's socialism.

3

The Computer in the Classroom

In the 1950s Butte, Montana, was a pretty rough mining town, hard wintered for hard people. In a place with forty-below Januaries, apartment houses had external stairs, and the major cause of winter death was freezing after a night in the bar. The town celebrated boxing , with semiorganized fistfights in bars run by former contenders gone not-too-soft. It allowed but did not celebrate whoring—Saturdays I delivered groceries to the ladies in the house on Mercury Street, a block from the high school and now a National Historic Place with a mannequin prostitute out front. The high school football field was gravel; grass wouldn't grow after August, and Astroturf had yet to be invented.

All in all, a tough place for nerds or for any curious kid. The town had a college, the Montana School of Mines, but it was an intimidating place, and high school students could not use the library. So for books it was the city library, in a wonderful wood-paneled building with ceilings a mile high and lights so dim you could make out the books in the stacks only in bright daylight. (The building burned to the ground around 1960; a few years later much of the rest of downtown was burned for the insurance money when the Anaconda Copper Company was bought by an oil company that stopped the only industry in town, digging copper ore.) Unfortunately, there weren't so many books in the library. Interested in philosophy, I could find only one book by a philosopher with a recognizable name, Schopenhauer's *Essays*. The essays were dismal drivel—Schopenhauer was a famous pessimist with a totalitarian streak—capped by a stupid denunciation of women. At sixteen, I thought women were about the most fascinating thing going.

Today, thanks to the Internet, any kid in Butte has access to just about every bit of information available to a Cal Tech professor. For information *access,* the computer has changed the world. No surprise that education would be changed as well, but not as much as it should be. What follows is my personal story of how that did and did not come about.

I learned to program in FORTRAN in 1963, and I just hated it. With great effort some trivial function would be programmed on a keypunch, and out would come a huge box of cards, which then had to be carried to the computer, from which a printout would eventually emerge. If you missed a semicolon somewhere, the program did not work and you had to start over—the whole thing. If you tripped with that bunch of cards, and I'm a tripper, you had to start over. When, in my graduate course in quantum statistical mechanics, the final assignment was a long calculation requiring a computer, I instead lined up all my friends and had them pass along the steps in the computation to the next in line. I'm told Richard Feynman did something like that on a more serious scale during the Manhattan Project. Even in 1969 programming wasn't fun, and 1969 was the year it occurred to me that if it were fun, or even endurable, it would be useful for education.

That was the year Princeton first admitted black undergraduates (yes, American institutional racism is that close behind us). I failed about half of them. It happened this way. I was assigned to teach introductory mathematical logic to a class of about seventy students. I chose a textbook by one of the most eminent logicians of the day, gave regular lectures at the blackboard, homework assignments, a midterm examination, and a final examination. The students had weekly tutorial sessions led by graduate students, who were also the graders for the course. One morning at the end of the semester my graders appeared in my office to tell me I had failed every black student in the course, all seven of them. I asked if I had failed any white students. Yes—one.

Not an hour later a student with a name like Taylor Harrison Taylor IV, a not-too-bright blond lad, appeared in my office, said I had failed him, and demanded a C, else he would have his father fire me. I believed (wrongly—Princeton was good about that sort of thing) that he could do that, but I went to high school with Evel (né Robert) Knievel, and so I am used to crashing. THT IV failed the course without my regrets, but I wondered why all of the black kids had failed as well. These had to be pretty smart kids, or they would not have been admitted; they had to be ambitious kids, or they would not have wanted to attend an Ivy; they had to be brave kids, or they would not have wanted to attend a snobbish, traditionally racist college, which was what Princeton was trying to cease to be in 1969. Their problem was me or my course or both.

University courses have been taught more or less in the same way since the Middle Ages. A supposed master lectures, students listen and take notes, their questions are addressed by assistants of the master outside the lecture, and sometimes the students talk to one another about the lectures. Maybe, like medieval methods of detecting witches, medieval methods of instruction could be improved on.

Suppose the black kids were less well prepared for abstract mathematics than the others in the class, except, of course, for THT IV. The course organization gave students little opportunity to substitute sweat for background; the lectures presumed they had followed everything up to the point of the lecture; if they hadn't, they would be lost. Or suppose they were less good than others at taking timed tests. The testing was one shot each, midterm and final, no do-overs, no mulligans. So, the next term I was assigned that course, I reorganized. I found a textbook that divided the material into a large number of very short, very specific chapters, each supplemented with a lot of fairly easy but not too boring exercises. For each chapter I wrote approximately fifteen short exams, any of which a prepared student could do in about fifteen minutes. I set up a room staffed for several hours every weekday by a graduate student or myself where students could come and take a test on a chapter. A test was immediately graded; if failed, the student could take another test on the same material after a two-day wait. No student took the very same test twice. I replaced formal lectures with question-and-answer sessions and extemporaneous minilectures on topics students said they found difficult. I met with each student for a few minutes every other week, chiefly to make sure they were keeping at it. Grading was entirely by how many chapter tests were *passed;* no penalty for failed tests. To receive a B, a student had to pass tests on the material I had covered the previous year; to receive an A, the student had to master considerably more than had been required for the same grade the previous year.

That semester I had about the same number of black students as I had had the previous semester, none of whom had taken the course before. Half of them received a B grade, the others received an A. I had discovered something.

What I had discovered was that a lot of capable students need information in small, well-organized doses, that they often do not know what they do not know until they take a test, that they need a way to recover from their errors, that they need immediate feedback, that they learn at different rates, that they need a chance to ask about what they are reading and trying to do, and that, if given the chance, motivated students can and will successfully substitute sweat for background. This turns out to be about all that education experts really know about college teaching, and of course it is knowledge that is systematically ignored.

I also discovered that it is not physically possible to teach that way: I was ready for the hospital by the end of term. (My efforts were not appreciated. I received the lowest teaching evaluations ever. The Princetonians thought the course was a ruse to save myself the trouble of formal lectures.) But it did occur to me that this sort of course was ideal for instruction by computer. The computer could be programmed with small modules, with interactive questions and prompts, with many penalty-free tests for each module, even with automated problem-solving aids. That would work for a lot of introductory mathematics and for science and engineering classes, and maybe even, if sufficiently cleverly done, for history and philosophy. But I was not about to try to create such a course in FORTRAN. I never taught logic again, anywhere.

Instruction by computer got started in small ways in the 1970s and has swelled with the development of the Internet. Internet delivery initially ran smack into bureaucracy. When Ken Ford, then a professor of computer science at the University of West Florida, first offered a course over the Internet subscribed to by students from all over the United States as well as by some foreign students, he was stopped by the requirement that all students enrolled in the Florida university system must have a set of vaccinations. The Florida Board of Education eventually agreed to virtual vaccinations for nonresidents. Nowadays university curricula are frequently delivered over the Internet without vaccinations, but that is not entirely a good thing.

Standard Internet courses deliver standard text material or videos of standard lectures. Some courses have online chat rooms or other live sources of help at scheduled times, and most have facilities for enrolled students to write back and forth to one another. The intent is to reproduce remotely as much of the standard medieval university situation as possible—read, hear lecture, ask a graduate assistant, talk with fellow students—which makes about as much sense as designing a chess-playing computer to imitate the performance of a woodpusher.

Beyond its use as a word processor, the computer in elementary and secondary school classrooms is typically a distraction programmed with minor exercises and spreadsheets. John Anderson, a psychologist who worked extensively on the development of "intelligent tutors" to teach geometry, algebra, and computer programming, gave up. Anderson and his colleagues developed a program to teach algebra to high school students, testing and refining it on real students until they could demonstrate that students learned more in less time than they did in conventional classrooms. Then they persuaded some public schools to make the computer program available in some algebra classes. The arrangement allowed students' behavior with the computer program to be automatically recorded. Anderson and company soon found out that the presence or absence of the computer programs made no difference to how much algebra was learned in real classrooms. Why not? Because the computer program did not closely follow the prescribed algebra curriculum in the prescribed sequence, so teachers would not use it. Too much extra effort.

Anderson and his students rewrote their software to conform to the syllabus and textbooks in the targeted classrooms. Again they waited for results, and again the results were disappointing. Students in the classrooms with the intelligent tutors used the software, but they did little better than students in the classrooms without the computer aids. The records of student activity showed why. The algebra curricula emphasized "real problems" and working with "real data." As anyone who has worked with real data knows, most of the time is consumed in managing and formatting files, and that's how the students' time was occupied. Very little time was left for learning or understanding algebraic formulas, algebraic properties, and their relationships. Anderson estimated that about twenty-five hours each semester were spent on algebra in these algebra classes. As he put it to me, "Nothing significant can be learned in twenty-five hours."

So back to college. In the late 1990s Joel Smith, then dean of library and information services at a California community college, and Richard Scheines at Carnegie Mellon teamed up to develop a computerized interactive course on causal and statistical reasoning that could be delivered over the Internet. The technology to make that feasible was just being developed, and it took their team several years and nearly a million dollars in real or in-kind funding to develop the course, which is now taught in several colleges and universities through Carnegie Mellon's Open Learning Initiative. The course uses the principles I had so painfully recognized twenty years before: short, specific modules that build on one another, written material that requires student responses as they read along, quizzes at the end of each module, immediate grading of quizzes, no penalty for failing a quiz except the requirement that another quiz on the same material be taken after review, progress to the next module allowed only after success on a quiz on previous modules, and occasional tests unifying previous material, with the same no-penalty conditions. Students can proceed at their own pace. Smith and Scheines added a final exam for real grading.

Recently the Andrew W. Mellon Foundation sponsored several investigations into how well and how cheaply computerized instruction delivered instruction. In randomized and nonrandomized comparisons, the Causal and Statistical Reasoning course, supplemented with tutorial help, delivered learning outcomes as good as or better than courses on the same material with lecture formats supplemented with tutorial sessions. The cost of computerized instruction—not counting the cost of development of the software or the cost of computers (which the participating institutions already had installed), was about two thirds the cost of conventional instruction. In another combined effort of the statistics and psychology departments at Carnegie Mellon, an introductory course on statistics using the same principles has been developed and tested. Again, learning results were as good as or better than with conventional lecturers, but the experiments found something else of interest: when the computerized course was combined with lectures, students could complete the course in half the time with comparable learning results.

In the 1990s the University of California recognized that it was (and is) facing a demographic landslide. Thousands of young people would be ready to enter colleges in the state system, which had no room for them. Richard Atkinson, then the president of the university and also a distinguished psychologist, organized a large meeting of faculty and administrators from across the state to discuss the problem. Addresses were given by faculty and administrators with ideas, and one of the ideas was presented by Patricia Kitcher, then a professor of philosophy at the University of California, San Diego. Kitcher's pitch was that with $100 million the University of California system could construct as many as fifty interactive, self-paced, computerized freshman- and sophomore-level courses in a wide variety of areas—mathematics, statistics, physics, chemistry, biology, history, social sciences, and so on—delivered over the Internet. The courses could be developed by faculty in collaboration with expert commercial computer companies, such as Pixar. Computer facilities for students could be made available in libraries, junior colleges, secondary schools. Much of the first two years of the university curriculum could be taken while students lived at home, followed by two or three years on campus. The cost to families and to the state would be enormously reduced, and the flow-through—the number of students who could complete a degree within four years on campus—would be increased by as much as 25 percent.

Of course, it did not happen. Instead, the University of California built a new campus in Merced, at a cost in the neighborhood of $1 billion. *Merced?* The Merced campus addressed political problems, made jobs for construction workers, and made money for developers. But that wasn't the entire reason. After the meeting, Atkinson told me it wasn't about the money, that $100 million was not very significant in the University of California budget. The problem, he said, was that if the money were made available, he believed that the California faculty would take it; he didn't believe they would produce the courses. Perhaps he was right, perhaps not: as I write, the California State University system is overwhelmed, short on money, and long on students. The federal Department of Education spends about $700 million a year on "educational research," which subsidizes work much worse than Professor Darling-Hammond's. The money is there. Find the right professors—they won't be in schools of education.

4

Cosmic Censorship

When I was fifteen, my wonderful, devoutly Catholic next-door neighbor, Rita Harrington, looked over the backyard fence and found me reading *The Origin of Species.* "That's a dangerous book," she told me, and of course she was right.

A lot of Americans, a lot of Europeans, and perhaps most of the people of the rest of the world do not believe in evolution. In particular, they do not believe Earth formed more than 4 billion years ago, they do not believe that species were altered and extinguished over time by natural genetic variation and natural selection, and they emphatically do not believe that our species evolved from an earlier hominid species and ultimately from very primitive single-celled life. They do believe that a god created the world and formed the species that live with us. Scientifically, a lot of the world, not just Kansas, Texas, and Oklahoma, is the Neolithic plus the pickup truck. Scientific rationality has fewer endorsements than freedom of speech (which, according to a majority of the nations voting in the United Nations Human Rights Council, should be outlawed).

For America, evolution is a problem about schools, almost always in this form: should teachers of biology be allowed to say that there is an alternative to the theory of evolution favored by many people—a few of them scientists—variously called creationism or intelligent design but really meaning: God did it. What freedom of speech should we allow to high school biology teachers? In many schools, Pittsburgh's, for example, a high school history teacher is free to instruct his students that Hurricane Katrina was caused by secret wave machines the federal government placed in the Gulf of Mexico and used to try to drive black people out of New Orleans, but a high school biology teacher is so severely censored that she is not permitted to discuss why special creation and intelligent design claims are not scientifically legitimate. In a real school, the scientific legitimacy of intelligent design would be the subject of a discussion led by a scientifically informed, philosophically sophisticated biology teacher. There are some of those, I am sure, but not enough, and I am equally sure that in many classrooms the biology teacher goes through the required evolution instruction and tells the students not to believe a word of it.

The community of people who trust science does not trust high school teachers to be scientifically informed and conceptually sophisticated, so its recourse—a practical, political recourse—is censorship. The deal American public schools try to make about anti-Darwinian claims is the same deal American public schools have made about religious claims: just don't talk about it. (Except maybe in Utah.) The two deals are actually one and the same.

Censorship may or may not be the best policy we can practice, but it is intellectually dishonest in its heart, destroys real curiosity, and fails. The kid who learns evolution in school goes home or to church and is told "They *make* your teacher say that, and they won't let her or you talk about God's Truth." The kid has no answer, even to herself, because she knows for a fact that the teacher is required not to talk about God's role or nonrole in creation or about God's Truth or God's Falsehood, either one. G. K. Chesterton once observed that there is no culture if there is no culture in the streets. Censorship in schools is bound to fail in a community that believes what is censored and denies what is not, and it does fail: only 15 percent of American adults believe that Darwin had it about right.

The intellectual dishonesty begins with history and its fundamental theorem: nobody reads. The paradigmatic work of modern physics, the creation work, the founding work, is Isaac Newton's *Principia;* much of its content—the three laws of motion, the elementary dynamics of the solar system—we teach in high school physics. We do not have students read the book (for which they should be thankful) or its last pages, which are about God and how Newton's system reveals his work. We pretend that evolution has nothing to do with religion, but students (and, I reckon, most biology teachers) do not read Darwin's letters or notebooks, which show, among other things, Darwin's struggle with the recognition that his theory cut the ground out from under the Christianity of the nineteenth century. The great debates among scientists and clerics of that century and the next over the moral and religious implications of Darwin's work are hidden from our students. The social and intellectual history of modern science is inseparable from religious issues, and considering why we ought to give credence to Darwin is inseparable from considering why we ought not to give credence to religiously motivated criticism or to the doctrines from which it springs.

Formulation of a general rule for separating science from nonscience has defied methodologists for a long time. Many philosophers and some statisticians think that rationality requires only that people's beliefs accord with the theory of probability and that they change their beliefs when they acquire new evidence according to a rule propounded by (the Reverend) Thomas Bayes in the eighteenth century. But propositions can be given arbitrary initial degrees of belief, including certainty, and according to Bayes's rule certainty cannot be rationally dislodged by any evidence whatsoever. The antievolutionary dogmatist is rational by that standard. For that reason and others, many methodologists reject the Bayesian standard. Karl Popper, one of the most influential philosophers of the twentieth century, argued that scientific claims are distinguished by being falsifiable—not false necessarily, but capable of being tested. Almost no one who thinks long about Popper's argument agrees, because science as we commonly recognize it is filled with claims that cannot be tested in isolation, only in large combinations or (as it seems with contemporary string theory) not at all. Newton's second law claims that the total force acting on a body is proportional to the mass of the body multiplied by its acceleration. Newton could calculate accelerations, but he had no mass meter and no force meter. His second law could be tested only in combination with other hypotheses that specify forces. But, antievolutionists may well ask, cannot combinations of the design hypotheses with other hypotheses be tested? The National Science Foundation's public defense is equally lame. It points out a few of the striking coincidences and patterns of distribution among species and insists that they are explained only by evolution. But the NFS provides no intelligible, clear, general account of what is an explanation and what is not. Why is "God so willed it" not an explanation? Perhaps rightly, antievolutionists will think they are being cudgeled.

A few years ago, two of my former students testified at a trial in Dover, Pennsylvania, concerning whether creationism could be taught in high school. One of them was a middling student who, after graduating, wrote an excellent book doggedly and carefully answering one creationist argument after another. The other was a smart, intellectually evasive student (he once demanded that a graduate course on postmodernist views of science be given; I offered to give one with texts he could choose—no readings appeared because, he explained, I would *criticize* what they said) who argued before the court that intelligent design had the social earmarks of science and that there is no general criterion demarcating science from nonscience. Both of them were right that far, but of course intelligent design isn't exactly science, or at least not good science.

Millions think or have thought that Freud was doing science, good science. I think not. We can argue the case using scientific standards: was there investigator bias; did Freud bully testimonies from his patients; how often did his therapies succeed and by what criteria of success; was Freud candid about the changes in his hypotheses over time and about the evidence he himself provided against his own theories; was there any independent testing of the reliability of dream interpretation. And so on. We can do the same with intelligent design and creationism. Intelligent design "research" consists in trying to punch holes in the case for evolution by natural selection: describe some really complex biological system and claim there is no way it could have evolved from simpler systems; deny assumptions in radioactive dating; point out gaps, or alleged gaps, in the fossil record and claim they will never be filled. There is really only one positive hypothesis in the enterprise: God did it. Intelligent designers claim that they say something weaker, that an Intelligence did it, but they really mean that God did it. But "God" can be replaced by "the Wicked Witch of the West" or "the Committee of Demons," and that will work as well—perhaps, in view of the state of the world, better—in any intelligent-design explanation of biological phenomena. God's will could be attached to any testable biological hypothesis; the combination would still be testable, but it would be just as testable if God's will were deleted. God adds nothing. Scientific hypotheses unify in many ways; for example, they show that some general quantities are forms of others, and the identifications transform some phenomena into others that, without the identification, would seem altogether separate. Copernicus identified features of unobserved planetary orbits with features of observed motions of the planets against the stars, and—lo!—patterns of observed motions known a thousand years before emerged as mathematical truths. Citing God's will does none of that. There is a lot more to say, both positively for evolution and negatively about intelligent design. Good biology teachers should be allowed to say it.

The real complaint of science education advocates ought to be with the system by which we train and evaluate teachers and set curricula. We leave training to schools of education, the bottom feeders of academe. We require of would-be science teachers no comprehension of the history of science, no sensibility of the reliabilist rationales of scientific methodology, no demonstrated capacity for abstract reasoning. We leave curricula to state boards of education too often peopled by dentists and education professors. Good states require that teachers show they know something of their subject, including evolution if they are biology teachers. That is less than they need to know, less than they need to think through, and far less than they should be able to talk about in a classroom.

II
The Environment

5

The Greatest Chemical Engineer
There Ever Was:
A Cautionary Tale

In 1989, shortly after the *Exxon Valdez* spilled more than 11 million gallons of oil off Alaska and just after the governor of Arizona was impeached, I read the newspapers for the winter of 1917. A bit late, sure, but the news from 1917 was interesting: a big oil spill at Port Arthur, and the governor of Arizona had just been impeached. There may be something new under the sun, but nothing having to do with oil or politics seems to go around only once.

Nineteen seventeen was a year of the automobile (and, of course, the White Sox). While the nation entered a war its president had promised to avoid, the Espionage Act empowered the government to ban from the mails any publication the administration disliked; Germans, feminists, socialists, Wobblies—pretty much any dissenters—were jailed; and the states of the union were preparing for the automobile. New Jersey established a highway commission that year, and so did Florida and Wisconsin. Michigan painted the first center stripe. The nation moved by Cadillac, Dodge, Olds, Reo (for Ransom Eli Olds), Ford, Packard. The clumsy, fragile horseless carriages of the previous decades had been replaced by faster, more powerful, more robust, mass-produced automobiles with gasoline—not steam or electric—engines, and they were selling much better than hotcakes.

The internal combustion engine needed gasoline, and for gasoline it needed oil. The first oil well was drilled in Titusville, Pennsylvania, in the middle of the nineteenth century; by 1917 oil was mined across the country, and oil refineries distilled the stuff into gasoline. Converting oil into gasoline required no fundamentally new chemical trick. Oil refining uses basically the same process used to separate whiskey from water and the same style of equipment more or less (actually, more)—the distillation column or still. By 1917 the spawn of Rockefeller's Standard Oil provided almost all the oil America wanted, and it wanted a lot, roughly 400 million barrels of the stuff each year and growing, a quantity George Otis Smith, the director of the United States Geological Survey, thought almost unimaginable and clearly unsustainable. In 1920 he published a think piece in *National Geographic*, worrying about where America would get oil when its own ran out. He was not alone in thinking that the supply of oil would soon be exhausted and that either some means had to be found to make the unknown but finite supply move automobiles farther or some alternative fuel had to be substituted. An alternative fuel was at hand, ethanol from corn. Calculations, unfortunately, argued that half of the entire American corn crop would be required to replace gasoline at then current levels of consumption and production. The America to come could eat or drive, but not both.

Thus in 1917 or soon thereafter, scientific leaders in the United States believed the nation to be in the very predicament that, with better justification, their heirs found in 2007. George Otis Smith had neither theory nor evidence that the United States was, in 1920, about to run out of oil. He simply thought the quantities of oil the nation required were preposterous. We since have seen correct predictions by petroleum geologists of when—nearly to the year—U.S. petroleum production would decline (the mid-1970s) and theoretically well-founded predictions—so far borne out, new discoveries notwithstanding—that world oil production would flatten early in the twenty-first century and then begin to decline. And the quantities are a lot more preposterous than they were in 1917: the annual oil production that Smith thought so unsustainable amounts to about five days of current world oil production. Then as now, the old chemistry of alcohol offered a conceivable solution, and others were sought in a new chemistry.

Every internal combustion engine has a compression stroke in which oxygen and gasoline vapor are compressed by a piston. Roughly, the greater the compression, the more energy is transferred to the piston upon combustion, and so the greater the horsepower of the engine. How much a gasoline/air mixture can be compressed before spontaneously exploding depends in part on the composition of the gasoline. "Knock," something unfamiliar to modern drivers, is the uneven or incomplete or premature combustion of compressed gasoline and air in an engine cylinder. In 1917 engines knocked a lot, and the phenomenon limited the useful compression in engines, and thus their horsepower. A 1917 Packard Model E mustered 35 horsepower. A Cadillac sedan had 31.25 horsepower with a compression ratio of 4.25 to 1. A 2006 Caddy had a 10.5 to 1 compression ratio.

Gasoline is a mixture of compounds of carbon and hydrogen called alkanes, a class of molecules characterized by the fact that between any two carbon atoms in an alkane molecule there is at most one chemical bond. Ethane, for example, has two carbon atoms, each bonded to the other, and six hydrogen atoms, three of them for each carbon atom. Ethanol is ethane in which one hydrogen-carbon bond is replaced by a bond from that carbon atom to an oxygen atom, which is in turn also bonded to a new hydrogen atom. Octanes are alkanes with eight carbon atoms and eighteen hydrogen atoms. Unlike ethane, octane has a variety of molecules that differ in their geometry and therefore in their physical and chemical properties. One of them, 2,2,4-trimethyl isooctane, can be compressed a lot without spontaneously exploding. So if petroleum is refined just right, and the right octane is extracted, high compression is possible—but a lot of the petroleum is then of no use in gasoline, and the aim of making a limited supply of oil last longer for transportation is defeated.

In 1917, then, there appeared to be just two options: use up half of the nation's corn supply or find some chemical means to eliminate knock; that is, to let gasoline with a high percentage of nonoctanes burn just as if it were high octane. Thomas Midgley, Jr., found a way to the second solution: tetraethyl lead.

Charles Kettering was no mean inventor—before World War I he had already devised and sold the electric starter for automobile motors—but his single greatest discovery was Midgley, the greatest chemical engineer there ever was. Midgley graduated from Cornell in 1911 as a mechanical engineer and went to work for Kettering's Dayton Engineering Laboratories Company. In 1916, during World War I, Kettering put him to work on the problem of engine knock and the use of ethanol to fuel airplanes. Midgley soon established that ethanol or a gasoline-ethanol mixture reduced knock, but at war's end he began to focus on a search for chemicals that could be added in small quantities to gasoline to eliminate knock. Kettering himself had experimented with adding iodine and dyes to gasoline. In 1916 Kettering's operation was taken over by General Motors, and Midgley, reverting to Kettering's earlier strategy, experimented with aniline dye additives to shore up fuel research under the new ownership. Kettering suggested adding selenium in 1921, and a variety of metal and other elemental additives were tried—bromine, sulfur, iodine, tin, selenium, tellurium, titanium, lead. Lead worked best, and near the end of the year it was discovered that drops of tetraethyl lead in gallons of gasoline sufficed to eliminate knock.

Ethane, when combined with mixtures of certain metals and sulfur, results in compounds in which four ethyl radicals are chemically bonded to a metal atom. In tetraethyl lead, of course, the metal atom is lead, with the ethyl groups splayed out as four bent legs of a tetrahedron.

With backing from Standard Oil, General Motors, and the DuPont Corporation, Kettering formed the Ethyl Corporation to manufacture tetraethyl lead, and lead additives began to appear in filling stations. Manufacturing and development plants were located in Dayton, Ohio, and Deepwater, New Jersey, and in 1924 in Bayway, New Jersey. Almost immediately things began to go very, very wrong.

By 1923, the year the first tank of leaded gasoline was produced for commercial use and the year the three top-finishing Indianapolis 500 cars ran on leaded gasoline, Midgley had already suffered a serious illness from working with tetraethyl lead. He had been warned by researchers around the world of the dangers of the substance. In Dayton two workers died and several were injured, but the events caused no public storm. Similar accidents occurred in Deepwater. Then, in 1924, a major accident occurred in the Bayway plant. Several workers went "insane," and eventually seventeen died and many more were injured. Newspapers from around the nation followed the story, not always with scientific accuracy. (The specific cause of the workers' deaths and injuries still seems uncertain. Newspapers attributed it to "loony gas" from the plant.) Midgley gave a public address in which he claimed that tetraethyl lead was harmless and, to prove as much, poured the stuff all over his hands—not a good idea, since tetraethyl lead is absorbed through the skin. General Motors arranged a review of the chemical threat with an intimidated U.S. Bureau of Mines in Pittsburgh. Eventually, under a contract that gave GM censorship power over publication, the bureau produced a glowing report—and GM formed an internal committee to review the safety of the stuff. Surprisingly, GM's internal committee returned a negative report on the safety of tetraethyl lead, with the result that, to no surprise, the report is still not in the public domain. Several states and cities (Pittsburgh among them) banned leaded gasoline. Most risky to the company interest, the federal Public Health Service called hearings on the safety of tetraethyl lead.

The hearings were held in 1925 while the Ethyl Corporation suspended sales of its product. In the course of a few hours academic experts testified that lead exhaust from gasoline posed a widespread threat to public health, while industry witnesses argued that accidents at plants could be avoided by improved chemical engineering and worker training and that any harm from low-level lead exposure in the wider environment was unproved. The industry was correct on the available evidence; but the academics, including Dr. Alice Hamilton, a toxicology expert who was the first female faculty member at Harvard, were right about the future. A committee appointed by the surgeon general in response to the hearings reported in 1926 that there were no grounds for banning tetraethyl lead. Ethyl was back, big time, and stayed back for fifty years.

Midgley was not done. Keeping food cold enough to delay spoiling is an old problem once solved by the rich in America—Thomas Jefferson, for example—by having slaves cut ice blocks in the winter and storing them in an icehouse for the rest of the year. An icehouse well insulated with sawdust would keep ice throughout the seasons, with the ice carried bit by bit to a food cabinet. Sans slavery, the method lasted in America until after World War II. (In the 1940s, each week, while into one house after another large men carried blocks of ice held by tongs on leather pads across their shoulders, I and my pals would climb into the ice truck and steal huge shards to lick and drip and swing like frozen swords.) By the 1950s just about every white family had a refrigerator, a device invented in the nineteenth century but made safe by Thomas Midgley only around 1930.

Refrigerators are simple, and they last almost forever. To make a refrigerator, fill a closed tube with a gas—the air kind of gas, not the fuel kind—with one segment of the tube exposed to the outside of the space to be cooled and another segment exposed to the inside of a room or closed box. Near the part connected to the outside, run the tube in and out of a compressor that reduces a volume of gas to a smaller volume and liquefies it. The phase transition under pressure from gas to liquid releases heat, which is radiated out of doors—energy has been passed from the fluid to the outside. Now let the liquid form a gas again as it gathers heat while passing through the part of the tube connected to the space to be cooled. Repeat the process, over and over.

Practical refrigerators were developed around 1860 using ammonia as the coolant, and into the 1930s almost all refrigerators compressed and expanded ammonia, with just one problem: ammonia leaks in closed spaces produced catastrophic fires and explosions. Ammonia, NH_3, combines with oxygen to form nitrogen oxide gas and water, and the reaction produces heat. Ammonia burns and, with enough concentration, burns explosively. The obvious solution was to find some other substance that efficiently absorbs and efficiently gives up heat, that does not burn or explode if leaked, and that does not corrode pipes carrying the coolant. Midgley found a solution, chlorofluorocarbons, composed of chains of carbon atoms with each carbon atom bound to chlorine or fluorine atoms. Midgley did not discover this class of compounds—they were old hat—but he found a use for them. Freon became the most common refrigerant, and chlorofluorocarbons were widely used as the propellant in aerosol cans of hair spray and shaving soap.

What gets manufactured gets into the environment, and chlorofluorocarbons were no exception. Rising to the stratosphere, the stable molecules hang around until struck by ultraviolet light, which is a lot more intense (350 million times more, roughly) up there than down here. UV light breaks off intensely reactive chlorine atoms and fluorine atoms from the chlorofluorocarbons. Each atom of free chlorine reacts with ozone, O_3, in the atmosphere, and it does so by catalysis: when the reaction is complete, the ozone is gone but the chlorine is still there to do it again. So Midgley's refrigerants began taking out the ozone layer. Ozone in the upper atmosphere is itself the major absorber of UV radiation, via a process that does not decrease ozone concentrations—the UV photon breaks the O_3 molecule into O_2 and free oxygen, and the free oxygen atom combines with another ordinary O_2 molecule to produce a new ozone molecule. So, as the ozone is depleted by chlorofluorocarbons, more UV radiation reaches the surface of Earth. And the UV that ozone would otherwise absorb is deadly.

Worldwide monitoring of the ozone layer began in the 1920s, and in the 1970s systematic decreases in ozone concentration in some regions—especially over the Antarctic—were found. In the 1980s the Montreal Protocol phased out the production of chlorofluorocarbons worldwide. So ended Midgley's second great contribution to human comfort. (The Montreal Protocol has often been cited as a model for international agreement about limiting carbon dioxide emissions, but that is a far-fetched hope: chlorofluorocarbons were a small part of the world economy for which substitutes were available, but everything we in the world do—everything—emits carbon dioxide.)

There is no evidence that Midgley meant anything malign in the application of his genius. His public defense of the safety of tetraethyl lead is perhaps most charitably understood as misplaced hope driven by pride of authorship, and there is nothing I know of to suggest that in his work on gasoline additives and refrigerants he meant to sacrifice anyone's health or life for profit or convenience. In the 1920s the environmental risks of tetraethyl lead were foreseen and predicted, but largely by one guess against another. In the 1930s, so far as I know, the environmental hazards of chlorofluorocarbons were not even guessed at. Nonetheless, the world took revenge on Midgley as if his life had been plotted by Sophocles: after contracting polio in 1941, Midgley, ever inventive, devised a system of ropes and pulleys to manipulate objects in his room. Entangled in the ropes, he strangled to death.

Kettering is remembered for the cancer institute he and Alfred Sloan, the boss of General Motors, endowed for the New York Cancer Hospital, later moved to land John D. Rockefeller had donated. Midgley is remembered by an annual prize in his name awarded by the American Chemical Society, but he should also be remembered as the poster boy of Unintended Consequences. All manner of technological schemes have recently been proposed for intervening in the environment on a large scale to stop global warming: blasting sulfur into the upper atmosphere, seeding the oceans with iron, putting huge mirrors in space, and so on. My advice about these would-be projects is this: remember Midgley.

6

Galileo in Pittsburgh

In 1633 Galileo was tried in Rome for promulgating heresies—most effectively in his *Dialogue Concerning the Two Chief World Systems*. He was convicted and kept under house arrest for the remainder of his life. Confined and watched, he still managed to write one of the most important books in the history of science, *Discourses and Mathematical Demonstrations Relating to Two New Sciences*, and have it smuggled to Holland for publication.

In the eyes of the Church, Galileo's fundamental error was one of *method*, his method of biblical interpretation. In the fifth century, Augustine (before he was canonized) considered how Scripture should be interpreted. Since many passages contradict one another or, if not one another then common sense, which are to be taken literally and which metaphorically? Augustine's answer, which became the Church's answer, was that the interests of the Catholic Church should be the guide. Galileo's answer was that science should be the Church's guide to truth and metaphor in the Bible.

Less famous trials of less famous scientists are common today in America, but the trials almost always concern whether the published work is fraudulent. Those accused risk losing their scientific careers entirely, and even risk criminal charges if the accusations are sustained. Usually the evidence of fraud is straightforward—data that were falsified, experiments that were reported but never done. But in

Pittsburgh there occurred a trial of a scientist that was, like Galileo's, fundamentally about scientific method. The superficial (but important) issue was the effect on children's intelligence of ingesting small amounts of lead.

From 1923 on, the Ethyl Corporation produced tetraethyl lead, refineries added it to gasoline, and American and European cars passed leaded wind out of their tailpipes. With the number of automobiles ever increasing (except during World War II), one would have expected lead to be newly dispersed throughout the environment. The Ethyl Corporation maintained it wasn't so: lead was naturally dispersed throughout the environment, and the additional contribution from leaded gasoline was insignificant. All that changed in the 1960s with the work of a chemical geologist, Clair Patterson.

Patterson, from Iowa, studied at the University of Chicago and worked there on the Manhattan Project during World War II. After the war ended, he completed a doctorate, focusing on the measurement of lead isotopes. As with other elements, atoms of lead may differ in their weight because, although they all have the same number of protons in their nuclei, they differ in the number of neutrons. Atoms of the same element—same number of protons—with the same number of neutrons constitute an *isotope* of that element. There are lots of different isotopes of lead, but most of them do not last very long. An atom of lead with atomic weight 181, for example, has a 50 percent chance of coming apart within 15-thousandths of a second. The two main uranium isotopes, U-235 and U-238, naturally decay to two distinct stable isotopes of lead, and the rates of decay are known. In principle, from the ratios of amounts of these two stable lead isotopes in a rock, its age can be calculated—provided that other sources of lead have not mixed in the rock. Meteorites seemed ideal. Because meteorites were presumably formed at or near the time the solar system formed and are protected from earthly contamination by the burning of their surfaces as they pass through Earth's atmosphere, the ratio of lead isotopes inside them should serve as a cosmological clock. Such a meteorite had been retrieved from Diablo Canyon, and Patterson extracted very small quantities of rock from it and separated the lead from other materials. The lead was put through a mass spectrometer—a device that ionizes atoms or molecules to give them an electric charge and passes them through a magnetic field, where the paths they follow are determined by their masses. His result, which is still the best estimate, is that Earth is about 4.5 billion years old.

Patterson went on to study lead concentrations in sediments of the world's oceans. He found that the recent rate of lead deposition had greatly increased over historical values, evidence that lead in the environment had greatly increased. The Ethyl Corporation's claims about lead in the environment were false and, by Patterson's estimates, wildly false. Patterson did not hesitate to say as much. He began to receive threats and offers of bribes and a degree of professional isolation. His opinion prevailed nonetheless, and by the 1970s the question was not whether lead had significantly increased in the environment but what harm it did, if any. Various researchers, including Claire Ernhart, a psychiatrist at Case Western Reserve University, undertook studies in the 1970s, and in 1979 a professor in the School of Public Health at Harvard, Herbert Needleman, and his collaborators published what came to be the most seminal and most controversial study of the effects of low-level lead exposure on the intelligence of children.

Needleman's plan was this: advertisements would be put around in Boston neighborhoods requesting children's baby teeth. The teeth would be tested for lead levels, and if testing and retesting of a child's teeth showed either unusually high or unusually low levels of lead, the mother would be asked to come in for an interview to provide information on the family situation: income, parent's education, and so on—altogether about forty factors that might conceivably have had an influence on the child's intelligence or on the child's exposure to lead or both—and the child would be given an IQ test.

The plan worked. Needleman's team collected more than 2,000 sets of teeth and interviewed about 240 mothers and children. After analyzing the data, Needleman's team concluded that exposure to low levels of lead lowers children's IQ scores a point or more and presumably therefore lowers their intelligence. (At least qualitatively, that conclusion has since been confirmed by any number of other studies.) Tetraethyl lead began to be phased out in the United States in the 1970s (because its byproducts destroyed catalytic converters) and hasn't been used here since 1986. It has vanished in other countries too, supposedly even in China, but it is still used to make aviation fuel more efficient—the problem that got Midgley into the mess to begin with.

Needleman and his collaborators reasoned from their measurements to their conclusion by using one of the oldest statistical procedures there is: linear regression. Linear regression is a method developed chiefly by an early-twentieth-century statistician, George Udny Yule, who, unlike some other famous statisticians of the time, was a nice guy. To estimate how much lead exposure influences IQ, one uses measures for lead exposure, IQ, and anything else that might be relevant for each person in a sample. (Ideally, the sample should be representative of whatever larger group of people one wants to generalize about, but epidemiology samples are almost never perfectly representative.) Then, holding all of the other variables constant (*statistically* constant—one doesn't actually do anything to the people in the sample except measure them), find a rule that *on average* best predicts subjects' measured IQ scores from their measured lead exposures. Then make a leap of faith: use the statistical predictions as estimates of the effects on IQ of a possible policy that would change lead exposure. Use the best prediction of how much IQ scores differ (on average) among members of the sample for each unit of difference in their lead exposures to predict how much the average IQ scores *would have changed* if the lead exposures of all people in the sample *had been changed* by a unit (while all other variables were forced to have the same value for all of the people in the sample.) What is calculated is a measure of association—the regression coefficient—between lead exposure and IQ, other measured variables held (statistically) constant. What is inferred is the average strength of a direct causal relation between lead exposure and IQ.

With the appearance of statistical programs for computers late in the twentieth century, regressions became a matter of pushing a computer key. Tests were developed early in that century for whether regression coefficients are "significant"—whether values that large or larger would be found in samples of the relevant size merely by chance even if the true value of the regression coefficient were zero. This was essentially Needleman's method (initially, he used a related method known as "analysis of variance"); it was standard at the time, and it still is; as we will see, it is fallacious except in special circumstances.

The regression coefficient will change depending on what other variables are also considered and held constant. The changes are complex and impossible to predict unless the causal relations among the variables is already known, which they are not. We would like to include the variables that are actually causes of IQ scores and leave out the other variables. The problem is how to know which variables those are.

Needleman tried at least six different "models" with different prediction variables, but his paper in 1979 reported only the model that best accounted for the total variation in IQ scores among the children in his sample. But six alternative models are trifling. With, say, forty measured variables in addition to lead exposure and IQ, there are (way) more than a trillion alternative choices for the set of variables, and many of these choices will give different values for the regression coefficient relating lead exposure and IQ.

And if these were not problems enough, there are many other difficulties that have to do with the misfit between regression and causality. For one thing, if A causes B and B causes C, regression to predict C will not find that A is a cause of C. Suppose, merely for illustration, that *family income* influences *child's IQ*, but only by influencing the amount of lead to which the child is exposed, which in turns influences *child's IQ*. In that case, regression that includes all of these variables will estimate that the effect of *family income* on *child's IQ* is . . . zero! The probabilities of children's IQ values would not depend on *family income* if *lead absorption* were kept constant statistically. If poverty causes children to live in hazardous environments that lower their IQ scores, that might be discovered by another study or another data analysis, but in this hypothetical situation it would not be discovered by regression using IQ as the outcome to be predicted. The late philosopher of science Hans Reichenbach called this kind of relationship "screening off." In our imaginary example, *lead absorption* screens off *family income* from *child's IQ*. Statisticians describe the same thing as "conditional independence": *family income* and *child's IQ* would be independent conditional on *lead absorption*. So one problem with regression is that it excludes indirect causes because of screening off.

There are larger problems. In many circumstances, regression will also include predictors that are not causes. Suppose, to exercise our imagination in another way, that *family income* has no influence on *child's IQ* but there is some unrecorded common cause of both. In this case the data would show that *family income* and *child's IQ* are correlated, and regression would wrongly estimate that *family income* influences *child's IQ*.

This kind of mistake about causal relations can happen even when two variables have no association, no correlation at all. Suppose A causes B, and there is an unrecorded common cause of B and of C, but neither A nor B causes C. Then variation in values of the unrecorded cause will produce an association of values of B and C, but there will be no association at all between A and C. Nonetheless, regression will find that A is a direct cause of C—because A and C will be correlated when B is controlled statistically, *conditionally* correlated.

For concreteness, imagine that *family income* influences *family education* and that some unrecorded variable influences both *family education* and *child's IQ*. In this circumstance, except by chance, there will be no association or correlation at all between *family income* and *child's IQ*, and yet regression will find a nonzero regression coefficient for *family income*—the estimate of the effect of changes in *family income* on *child's IQ* will be positive or negative but not zero, even though there is no causal connection and no association! If the regression coefficients are thought of as estimates of strength of causal influence on *child's IQ*, the inferences and conclusion will be wrong. The reason shows just how intricate the relations between probability and causality can be. There will be no correlation between *family income* and *child's IQ* because there is no causal pathway connecting them; *family income* is not a cause of *child's IQ*, and there is no common cause of *family income* and *child's IQ*. Why then, in this imaginary case, will Yule's method say nonetheless that *family income* does cause *child's IQ*? Because the method makes a mistake: in estimating the influence of *family income* on *child's IQ*, Yule's method looks at the association of *family income* and *child's IQ* when all other variables in the regression equation are constant—that is, it looks at the association of *family income* and *child's IQ* in each collection of families within which all the other variables have fixed values. So, in assessing the relation between *family income* and *child's IQ*, it looks at their association within collections of families in which *family education* is the same in all families. In our hypothetical scenario, *family education* is influenced by both *family income* and the *unknown causes* that also influence *child's IQ*. A conditional (on *family education*) association between *family income and child's IQ* necessarily results. With the methods that Yule and Needleman (and tens of thousands of other researchers) used, two causes of a third variable are always, necessarily, correlated—not independent—if the value of the third variable is held constant statistically. Regression makes a mistake in this hypothetical case. We will see that it probably made the same mistake with Needleman's data. Of course, none of these problems would arise if investigators knew the true causal structure to start with, but in Needleman's study, as in most others, that is exactly what was not known, exactly what had to be discovered.

Examples like these are easy to work out with some algebra, but various problems with regression (especially the last one above) were not widely understood until the 1990s and still are not as widely understood as they should be. Needleman's analytic techniques were more or less standard at the time, but they were risky business. Expert statisticians reviewing the conclusions in Needleman's paper made it riskier still.

In the 1970s, the decade of Needleman's work, statisticians produced a raft of criteria for selecting variables, and they described computerized procedures for selecting variables using some of these criteria. Early in the 1980s a panel of statisticians selected by the federal government reviewed Needleman's conclusions and recommended that he reanalyze his data using one of these techniques, *stepwise regression,* and Needleman did so. Stepwise regression is a method of selecting a subset of variables to use in a regression calculation. The method uses a "fitting function," which is essentially a measure of how much of the variation of the outcome variable in a sample—children's

IQ scores in Needleman's data—is accounted for by using a set of variables, but with the measure discounted or "penalized" by the number of prediction variables used. "Forward" stepwise regression starts with no prediction variables and adds the single variable that produces the best "fit" as measured by the fitting function. Then the second variable is added that most improves the fit, and so on, until adding further variables no longer increases fit. "Backward" stepwise regression starts with all of the possible prediction variables—all forty-some variables in Needleman's data—and removes the variable whose deletion from the set of predictors most improves the fit, then the next, and so on, until no further removals improve the fit. Stepwise regression can be run both ways, forward and backward, starting in either direction. There is no proof that any of these methods find the prediction variables that actually influence the outcome variable. The methods are "heuristic," meaning they are just guesses that in practice people have found useful, if not necessarily correct.

Needleman's backward stepwise regression reduced his forty-some prediction variables to only six, including *lead exposure* as one predictor. When regression values were then obtained, all six of the variables had meaningful regression coefficients—numbers sufficiently large that, in view of the sample size, they were unlikely if the true proportional constants were actually zero. The estimate of the coefficient for lead was approximately the same value as Needleman had previously obtained, and, as we will see, it was probably wrong. Then things got worse.

Sandra Scarr is—or was; she is retired at this writing—a psychologist well known for her work on child development. In 1983 the Environmental Protection Agency appointed her to an expert committee charged with reviewing the scientific literature on the effects of low-level lead exposure. The committee examined, among other studies, the publications of Claire Ernhart and Herbert Needleman. Scarr came to doubt that Needleman's data and analyses warranted his conclusions. Her concerns were not founded on any of the difficulties with regression methods that I have described above but on the methodological principles that statisticians and epidemiologists believed at the time. In 1990 Scarr and Ernhart were separately hired as expert consultants for defendants in a Superfund lawsuit, *United States versus Sharon Steel Company*. As experts in a suit in which Needleman's study was used as evidence by the government, Scarr and Ernhart were able to examine Needleman's raw data—not, however, under circumstances that allowed them to reanalyze his data by choosing subjects differently or by running regressions with different prediction variables. Their examination led them to file a complaint with the Office of Scientific Integrity, part of the National Institutes of Health, alleging that Needleman's work was "contrary to accepted practice" in the selection of variables for the regression, in the selection of subjects, in failure to use "multiple comparison" techniques, and in failing to report results for some outcome variables—subtests of the battery of IQ tests that Needleman used—that did not support Needleman's conclusion. Their complaint led to a formal inquiry—essentially a trial—by the University of Pittsburgh, where Needleman then worked, as to whether he had perpetrated a scientific fraud. Backed by the University of Pittsburgh faculty, Needleman took the extraordinary but entirely proper step of demanding that the hearing be open to the public. As a result, the Needleman case is one of the best-documented modern examples we have of misguided canons of scientific ethics based on misconceptions about sound scientific method.

After receiving reports from statistical experts who examined Needleman's data, his research notebooks, and his published papers, the hearing board rejected allegations of fraud against Needleman but severely criticized his work on three grounds: the way his subjects were selected, the way he reported that the subjects were selected, and the fact that he ran six regressions with different variable sets but published the results of only one of them. They explicitly did not criticize him for his choice of prediction variables, which the report said were a matter of judgment—meaning that the statistical experts whom the hearing board consulted had no definite idea how variables should be selected.

The hearing board's criticism of Needleman's way of sampling was correct, but it did not undermine the qualitative conclusion that lead exposure reduces intelligence, only how much and with how much uncertainty. The board rejected the suggestion that Needleman had selected his sample by considering both the children's lead exposure and their IQ scores. That would have invalidated even a qualitative inference that lead exposure influences IQ scores because in that case *lead exposure* and *child's IQ* would both have been causes of *membership in the sample*. Define a variable *sampled* that, for each child in the world, has the value 1 if that child was among the children Needleman sampled and the value 0 otherwise. Suppose *lead exposure* and *child's IQ* had each influenced whether a child received the value 1 for *sampled*. Then, even if *lead exposure* and *child's IQ* had zero correlation in the population as a whole, they would be correlated in the subpopulation of children for whom the value of *sampled* was 1.

In carrying out a regression, one is always implicitly conditioning on being in the sample—and we have already seen that conditioning on a common effect of two variables creates an association between them (even in nonlinear systems). So if Needleman had let (which he did not) both measured values of *lead exposure* and *child's IQ* influence his selection of which children to include in his sample for statistical analysis, he would probably have found an association between *lead exposure* and *child's IQ* even if the true effect were zero.

If the complaints about Needleman's sampling were sound in theory, the complaint that he had not published all of his original regressions—which was part of the point of Ernhart and Scarr's complaint that Needleman had not used "multiple comparison" techniques—was both accurate and nutty, a kind of nuttiness that is still commonplace in applied science.

The thought behind the complaint is this: suppose a scientist considers a hundred arbitrary variables, say the positions of the planets when a child is born, the heights of the child's grandparents, the house number of the child's address, and so on. Suppose that all of these variables would be *uncorrelated* with IQ if they were measured for *all* children, but the scientist has only a small sample of children. Now suppose that for each variable the scientist carries out a statistical test of the hypothesis that the correlation of that variable with IQ is zero. The test has the following guarantee: the probability is .05—one in twenty—that the test will wrongly reject a true hypothesis that the correlation of IQ and a particular variable in the population is zero.

Since there are a hundred variables, the scientist does a hundred tests. So (if the tests are independent) we should expect that for about five variables the tests will say the correlation is not zero when it really is zero. If a scientist does a hundred such tests and publishes only the tests that reject a zero correlation, we may be misled into thinking there are real correlations when there actually are none.

So there is a puzzle. Suppose with our tests we expect to be mistaken about 5 percent of the time in rejecting the hypotheses of zero correlation *if* there is no real correlation with IQ for any of the one hundred variables. There may be some variables among the one hundred, say V49, to give an imaginary name, for which the data (actually a statistical feature of the data, but the details are unnecessary) show an apparent correlation with IQ that is very improbable if the true correlation is zero. Which probability is the one we should rely on? Should we think:

> If all of the correlations were actually zero, then we would expect about 5 percent of the time to find a test result that rejects a true zero hypothesis; we found one zero-correlation hypothesis that is rejected. Therefore we should not give any credence to the hypothesis that the V49 correlation with IQ is different from zero.

Or should we think:

> Regardless of what was done with tests of other correlations, the hypothesis of zero correlation for V49 was rejected by a test that would reject a true zero-correlation hypothesis only 5 percent of the time on samples of the size in the data. So we should reject the hypothesis that the V49 correlation with IQ is zero.

Here is one way to see the strangeness of the first argument. The argument would work as well (or as badly) if there were a hundred scientists each testing (statistically) a different pair of the variables on the same data set. It doesn't matter to the probabilities who does the statistical tests. In fact, the first argument would work exactly the same way if each of the hundred scientists had her own separate, random sample of the same size from the same larger population and each of the scientists tested a single correlation hypothesis, a distinct one for each scientist. If the population correlations of all of the variables were really zero, by chance we would expect about 5 percent of the scientists to find a "statistically significant" nonzero correlation in their respective data, all drawn independently from the population. So if the first argument is accepted, we should severely limit how many studies there can be of different influences on a particular variable, children's IQ scores in this case, because the more studies there are with different variables, the more probable it is that a nonzero correlation would be found by chance. The order in which the studies occur is irrelevant, so we should also limit future studies. Science must be censored![1]

What about the complaint that Needleman did not publish all six of his regressions and their results? It would be a rare thing for a scientist to publish every hypothesis that passed through her head. Suppose it were required that scientists publish, along with the conclusion they advocate, every hypothesis they have confronted with the data in the course of reaching their conclusion. Then Needleman could not have published the result of the stepwise regression he was urged to produce by a previous panel reviewing his work. The computer that carries out a backward stepwise regression that settles on six variables out of forty tests a large number of hypotheses. No one should expect researchers to publish every such hypothesis, only the data and the identity of the computer program used in the search for a model. In many cases the computer program is the unfathomable procedure in the scientist's head as she mulls over alternatives and does back-of-the-envelope calculations. Einstein published several theories of gravitation before publishing the general theory of relativity in the autumn

1. The statistical literature has an answer, the "Bonferroni adjustment," which in my made-up example says that if we are testing one hundred hypotheses, we should not use the conventional .05 probability to reject any one zero-correlation hypothesis. Instead, with one hundred possible correlations, a hypothesis that a particular correlation is zero can be rejected only if, were that particular hypothesis true, the probability of the sample data test statistic would be less than (.05)/100. That's a pretty good guarantee that no causal connections would ever be discovered. More recent statistics uses better methods, for example the false discovery rate, which I leave to the reader's curiosity.

of 1915. We have no certain idea how many theories he considered and discarded along the way without publishing. Our confidence in general relativity does not require that we know any such thing and did not require it in 1915.

Needleman did what almost every good social scientist usually does and ought to do: he investigated which combinations of variables best explained his data. There was one important, fair criticism about Needleman's use of his data: he should have published it. Needleman had long refused to publish his data, claiming that lead interests would only use it to muddy the waters, but he did give the correlations of his final seven variables to two economists at Carnegie Mellon University, Mark Kamlet and Steve Klepper. They muddied the waters.

Kamlet and Klepper reasoned, plausibly, that there was some error in the measurement of the prediction variables—or that the actual measured numbers were of *effects* of the true causes of *child's IQ* together with measurement error. In models of this kind—which economists sometimes call "errors in variables" models—the strength of the effect of *lead exposure* on *child's IQ* cannot be determined. Economists say the parameter representing that influence is "unidentifiable," which means that a whole continuum of values is consistent with the model and with (almost) any correlations for the variables Needleman settled on. If that were not bad enough for Needleman's conclusion, the economists brought to bear a mathematical theorem of Klepper's. If a definite value of the error in measurement is assumed, the possible values of the influence are limited. Klepper and Kamlet showed that so long as the error in measurement was greater than about 10 percent of the recorded values, a value of zero for the influence of *lead exposure* on *child's IQ* could not be excluded. Maybe low-level lead exposure did not damage children's intelligence at all! The world had condemned low-level lead exposure too soon, on insufficient evidence.

Or maybe not. We have seen some of the mistakes that regression can make—it can even result in postulating that one variable causes another when there is no correlation at all between the variables. Is there some method that would show *only* the causal relations the data support? There are such methods. Richard Scheines, a professor of both philosophy and machine learning at Carnegie Mellon, helped to develop the first such procedures. A close friend of Klepper's, he applied one of these correct procedures to Needleman's correlations. What Scheines's computer program found was this: based on the correlations, three of Needleman's variables could have *no direct effect* on children's IQ scores. Two of the three variables, *mother's age at birth of child* and *father's age at birth of child*, have *no* significant correlation with *child's IQ*, and a third variable, *number of live births*, has no association with *child's IQ* when the value of *mother's education* is (again, statistically) constant—*mother's education* screens off *child's IQ* from *number of live births*. The significance of these facts was missed by Needleman, his critics and statistical examiners, and the two economists. When the irrelevant variables are thrown out, the identification problem is still not solved. Traditional statistical methods still cannot estimate the influence of *lead exposure* on *child's IQ*. But there is another way, Bayesian statistics, and Scheines made good use of it.

In "traditional" statistics, the kind of statistics that dominated in most of the twentieth century, no probabilities are assigned to hypotheses. Instead, hypotheses determine probabilities for data samples. In Bayesian statistics, probabilities are assigned to *hypotheses* as well. So there is a probability for a hypothesis and a probability for the data, given the assumption that the hypothesis is true. From these two numbers a third can be computed: the probability of the hypothesis *given* the data.

Easy to say but hard to compute. Bayesian statistics was mostly a toy idea until digital computers began to be widely available in the 1960s, but now its methods are widely used. The real question is where the initial or "prior" probabilities come from, since different prior probabilities will lead to different probabilities given the data. Generally they are just made up by assuming easy-to-compute probabilities and consulting experts.

Scheines assumed a bell-shaped prior probability distribution for the influence of *lead exposure* on *child's IQ* and also for the other parameters—a Normal distribution—and obtained prior estimates of how spread out it is (technically, its variance) from Needleman. Then he computed the posterior distribution. The mode of the posterior distribution—the most probable value for the reduction in IQ scores for every small unit of increase in lead in teeth—turns out to be about twice Needleman's estimate. Scheines went on to show that the result is nearly the same for other prior probability assumptions close to his.

Needleman is still at work and widely recognized for his contributions on lead pollution and its effects. Fox News occasionally vilifies him with selective quotations from the various reports and judgments on his work—even comparing his exculpation by the hearing board with the acquittal of O. J. Simpson. (Fair and Balanced, not so much; Frothing and Raving Mad, more so.) Scarr and Ernhart are sometimes dismissed as tools of the lead industry, but I know of no evidence that they were other than sincere. On the record, I think they were wrong; but I also think some of the hearing board's assessments agreeing with their technical complaints were also wrong, and Scarr and Ernhart could not be expected to have had better methodological judgment than the hearing board's experts.

It's a long saga with lots of morals. Impossible problems are solved only in fairy stories, so it is best not to create them in real life. If the goal is to get to the truth about some matter, it is certainly best not to create methodological strictures that hinder scientists from coming to know it. In the case of lead in gasoline, conceptions about scientific method and scientific ethics created just such a conundrum, and those conceptions continue to interfere with the search for truth in many scientific topics. One simple moral, still not always followed, is to make the data publicly available. It is easier to do that now, with the Internet, than it was when Needleman published his work. (But NASA, for example, still has not made its extensive data on aviation safety public in a usable form—and just try to get the data on space shuttle maintenance!) Another moral is that applied statistics has a lot of trouble with causal inference and tends to rely on methods that can be fallacious. The reasons for the widespread use of regression methods have to do with tradition, ease of use, ignorance of more reliable methods, and the need to produce results. Studies that conclude "can't say much" aren't publishable, but that is what more reliable methods of causal inference often say when applied to social and epidemiological data. A last moral is that while official scientific methodology in applied science is sometimes silly, real practice is often more sensible. There is something deeply troubling about the fact that applied science sustains that difference between rhetoric and practice. And finally, justice is not done when sensible scientists are tried by untenable standards, any more than it was in the seventeenth century.

7

Bert's Buick:
A Conversation on Climate Change

Bert, my late, consummately American father-in-law, was a small, smart man who admired Ronald Reagan and loved his family and big Buicks. Until he retired, Bert negotiated federal contracts for Westinghouse Power; his greatest disappointment came when he could not accept the position Reagan offered him as undersecretary of energy: the federal conflict-of-interest laws would have prevented him from going back to work in the only profession he knew. He won a Bronze Star and a Purple Heart (although that's not exactly winning) in World War II, hated the army, left churchgoing to his wife, and, with her, raised five daughters without a single premarital pregnancy. He did not keep company or friendship with Hispanics or blacks, but as is true of many middle-class white Americans, his antipathy was fundamentally cultural: he cared nothing for skin color or ancestry except as statistical indicators of what he regarded as acceptable behavior. He did not care much for the Catholic Church either, but he happily saw two of his daughters marry Catholic boys. He had been a poor guy with working-class jobs, and he was mostly but not entirely antiunion. Bert read, knew a lot of inconvenient facts, and brightened family dinner parties for me with hours of political arguments. After his retirement he read more and was a bit less Republican and a bit more cynical. Bert is my stereotype of the intelligent American nationalist conservative, a stand-in for millions of Americans whose voices are silenced by the noise of the professional right. This is a more coherent synthesis (with some liberties) of bits of conversation we had about climate change before and after the family dinners. He was occupied at dinner with his liberal daughters' teasing.

I would give Bert a ten-minute lecture like the following, and the truth of the matter is he would let me, and he would listen.

"Here's the synopsis. The United Nations Intergovernmental Panel on Climate Change reports bleak news: an estimated 4- to 5-degree Fahrenheit increase in average global temperature over the next century with a variety of unwelcome local changes in train and a 90 percent chance that the increase is due to human activities that have increased atmospheric 'greenhouse' gases.

"Between the 1850s and 2002 the temperature increased about 0.8 degrees centigrade, which is only about 1.4 degrees Fahrenheit, but most of that increase came in the last thirty years. The temperature varied around a fairly constant mean between 1860 and 1900, declined between 1900 and 1912 or thereabouts, then underwent a sharp increase between the two world wars. Another sharp decline in temperature occurred simultaneously with World War II, followed by a modest increase in temperature until around 1970 or so, when the temperature began to sharply increase again. It is tempting to link the changes in temperature with world economic behavior, but the period between 1930 and 1940 marks a worldwide economic depression, and the temperature nonetheless increased sharply over the decade. It is hard to say just what began to happen around 1970 to change things, but it seems plausible that what happened were a lot more people and a lot more wealth, notably in India, China, the United States, and Europe. China's real burst in gross domestic product (GDP) began with market reforms in 1978, with more than a tenfold increase by 2005. India's GDP increased dramatically between 1965 and 1970; by 1975 it had almost tripled, and by 2005 it was about fifteen times the 1965 GDP.

"China and India were not alone. The United States and Europe, and collectively the world, saw enormous economic growth as well. The nominal U.S. GDP increased about thirteenfold between 1968 and 2005. (Adjusted for inflation, the increases are smaller but still dramatic: one hundred dollars in 2005 would have bought about what fifty dollars would have in 1970—not that you could make the trade.)

"Making and distributing more stuff means emitting more carbon dioxide, a long-lasting greenhouse gas, as well as other chemicals. Today U.S. carbon emissions are about the same as China's, almost six times India's, and of course the per capita emissions from the United States are the highest in the world.

"There are two ways of predicting the future temperature of Earth; call them the physical and the statistical. The physical methods use computer models of the known—or at least believed—physics of the sun, the atmosphere, the ocean, and their chemical constituents and interactions. The computer models start with unspecified values for some key parameters and use some of the historical data to estimate their values. With the parameters estimated, the computer models can be tested by how well they fit the rest of the history, not just of temperature but of a variety of other climate factors: cloud cover, humidity, evaporation, and so on. There are at least eight different computer models of this sort; the best known are the Hadley Centre forecasts. The projected average temperature increases for Earth's surface by the end of the twenty-first century vary from just over 2 degrees centigrade to 5 degrees centigrade, which is the Hadley forecast. So a temperature increase somewhere between 4 and 9 degrees Fahrenheit is predicted between now and the end of the century.

"The statistical method fits the historical data to a set of equations specifying the temperature at any time as a function of previous temperature values, other measured variables, and chance variation. The simplest such projection is a 'least squares' straight line—the line that minimizes the sum, over all the years considered, of the square of the difference between the value on the line for each year and the measured temperature for that year, from the late nineteenth century through 2006. That line gives a projected value of the global temperature in 2100 somewhere between the projections given by the computer climate models. But the estimate of the coming temperature increase is probably *way* too low if recent trends continue. Aris Spanos, an economist at Virginia Tech, has pointed out that statistical variations in the data are not at all consistent with a straight-line projection. There are breaks in the trends of temperatures, with a dramatic increase between the world wars and another dramatic increase starting around 1970. There simply is no good statistical model of the entire series (although with enough parameters one could 'fit' anything). Fitting the data since 1968 yields a temperature increase at the end of this century of around 20 degrees Fahrenheit!

"The increases are in global averages, not local temperatures, but were the increase to hold across the United States, Pittsburgh would become Phoenix and Phoenix would become Hell. Something of the same may be true for carbon dioxide. In 1988 MIT economists reanalyzed the historical data and forecast higher carbon dioxide in the atmosphere in 2050 than did the global climate models.

"There is a kind of expert tension between the two ways of forecasting, the physical computer models and the statistical methods. Climate physicists understandably prefer doing and using the research that goes into the physical models. Statistical modelers reasonably object that good statistical tests of the multiparameter global climate models are difficult to come by, but there are good statistical models of time sections of the data. The statistical modelers can point to the history of economic modeling—big, complex computer models of national economies turned out to forecast less well than simple statistical models. But the physicists can reply that physical knowledge is more secure than economic theory, and the physical models have one huge advantage: they can yield detailed forecasts of *local* effects.

"The logic of the global warming argument is simple and persuasive: (1) we know that carbon dioxide, methane, sulfur dioxide, nitrous oxide, ozone, and other chemicals in the upper atmosphere can influence both the total radiation that reaches Earth and the balance between the energy of solar radiation that reaches the surface of Earth and the energy radiated back into space from Earth—changes in that balance, together with the total radiation received at Earth, determine whether Earth warms up or cools down; (2) we know that these chemicals, if produced on the surface, rise to the upper atmosphere, and some of them stay there for a long time—about a century for carbon dioxide molecules; (3) we know that production of these gases by human activities has increased enormously in the last fifty years, and we have pretty good evidence that it has increased over a much longer period; (4) careful measurement of other possible factors—variation in solar luminosity, carbon emissions from ocean sources, volcanoes, increases in water vapor, and so on—indicates that these factors are neither sufficient in size nor systematic enough in their changes to account for the pattern of atmospheric carbon dioxide increase."

Bert would reply:

"First, the '90 percent confidence' is 100 percent nonsense: individuals, not groups, have confidences, and there is no rational way to form a rational group confidence from a bunch of different opinions. Science by voting, which is what the IPCC did, is absurd. Besides, the measurements are uncertain and there could be some overlooked factors.

"Anyway, most of the upper atmosphere's greenhouse gas is water vapor, and water vapor accounts for the biggest segment of 'greenhouse' warmth of Earth, so carbon dioxide can't be that important."

"Yeah," I would say, "but the *increases* of temperature are due to *increases* of greenhouse gases, and the water vapor in the upper atmosphere, while increasing since 1980, isn't increasing much (although increases in other greenhouse gases will indirectly cause increases in stratospheric water vapor); and what people do (except for flying in high-altitude airplanes) doesn't have much direct effect on upper-atmospheric water vapor.

"So, Bert, you are up against both Isaac Newton (postulate no more causes than such as are true and sufficient to explain the phenomena) and Sherlock Holmes (when all other alternatives have been eliminated, what remains must be the explanation). Climate science conclusions about global warming and its causes could be all wrong; it could be that the temperature increases are due almost entirely to natural factors we have misestimated or missed altogether, and human production has not much to do with it. But a rational conservative should go with the odds, and the odds are that human-produced global warming is a Big Deal. My bet is that almost all of those who doubt global warming also doubt evolution.

Says Bert: "What about that famous physicist, Freeman Dyson, who doesn't believe this stuff? He says we don't understand enough about the biosphere to be confident of any predictions. And I am sure he believes in evolution."

"I said *almost all*, Bert. Dyson argues that we do not know a lot of stuff about how the ecosystem works or about how future practices in land management will influence climate. That's right, but we make rational and correct predictions all the time with very limited knowledge of all of the possibly relevant mechanisms. Nineteenth-century chemists did not properly understand atomic structure, but they made all sorts of correct predictions about how chemicals would combine, even about the existence of new elements. Nobody understands every detail of the motives people have for moving to cities, but demographers correctly predicted urbanization. We get by very well with uncertain and incomplete knowledge. Sometimes we are wrong, but surely it's rational to go with the best evidence we have.

"If the Hadley Centre projections are correct, much of the United States will be on average about 9 degrees Fahrenheit warmer by the end of the century, the Amazon will become so hot it will change in unpredictable ways, and northern Africa, already pressed for water, will be pressed further. Central China and India, the former already seeing expanding desert and the latter already at the limit of water supplies, will heat up substantially. A lot of the Arctic and some of the Antarctic ice will melt, and when ice on rock melts, the sea levels rise (and seas will rise, too, just because they are they are warmer; they have been rising slowly throughout the last century in any case). The Midwest will heat up. If sea temperatures rise, we can expect more or at least bigger major storms. The world will lose more species. And if Spanos is correct, things could be a lot worse than that. Your grandchildren and mine will be around then, if they're lucky, and suffer the consequences."

Even so, Bert would argue: "Global warming might leave those of us you and I really care about better off or at least well enough off. An American or Canadian might trade a northern sea passage in the summer for losing some polar bears, maybe some low-slung islands, an extra hurricane or two, and some Florida coastline. Anyone I know will never see a polar bear anyway, except in a zoo, so the worry about them is nothing more than artsy or sentimental. Environmentalists think it's *nice* to have all those polar bears roaming around, but I think it's *nicer* to drive a Buick. If fewer polar bears means that a lot more seals survive and eat a lot more fish, well, if things get out of hand we can kill a few more seals. The real problems are for all those 'stan countries in central Asia, for China and India, for the Latinos, for Africa, and for the Eskimos. We might get a few more mosquitoes, but why should Americans worry?"

"We should worry and more," I would reply, "because we are obliged to. We Americans generated and generate more air pollutants per person than any other nation of people on Earth. We generated and do generate all that stuff in order to make our lives better, but others will suffer the side effects. If your neighbor messes up your yard fancying up his own, he is responsible for cleaning up the mess. That's kindergarten morals. So not only should we cut back on our own carbon dioxide generation, we should give poor countries the means to do the same. Besides that, there is a universal human obligation to help those who are, or would become, the worst off among us and who are helpless to improve their circumstances by themselves. If we examine things from a disinterested point of view—if we supposed we did not know what condition on Earth you and I would be in, whether we would be comfortable middle-class men in America or starving children in Ethiopia—we would surely choose to have the world ruled by a policy in which the better-off give to the worst-off."

Bert would make some counterarguments: "Whatever pollution we Americans generated, we started real industrial expansion after World War II when we were rebuilding the world; if we hadn't done it, the world would be a lot poorer. Besides, why count how much pollution a country generates per person and not consider how many people the country generates and what it does to its own natural resources, its water and woodlands and grasslands? People act as a group, as a culture. Industry and transportation and the use of natural resources are collective cultural phenomena, and any ethical responsibility is likewise collective. The United States population—if you don't count immigrants and the illegal Mexicans here—has been constant for a while and is now actually falling. But Africa's population has exploded since

World War II and is closing on a billion people; the population is growing at almost 3 percent a year, the highest growth rate of any continent. If Africans are running out of water and soil and trees and will be unable to take measures against global warming, we Americans sure didn't do it to them. We had no colonies; all we did was buy what they would sell us. Sure we bought from thugs and dictators, but we didn't put the thugs and dictators in power; the Africans did when the European colonial powers left. If Americans had a choice about whether to produce more rather than less, then Africans had a choice as to whether to breed more or less and whether to tolerate good governments or bad ones. And China and India? China is passing us in carbon dioxide emissions, and the fact that they generate less per person than we do is irrelevant—the history of China is irresponsible human reproduction. Not our job to make up for the fact that the Chinese liked sex so much. Same for India, which was the craziest, most unjust, most inefficient society humans ever conceived outside of North Korea and maybe Burma and Saudi Arabia.

"Moreover, the idea of a universal human obligation to help the poorest in the world is nonsense. Just about nobody, except maybe the poorest, if anybody asked them, and a few liberals, thinks any such thing. For example, a few years ago a mountain lion killed a jogger in California, and then the lion was hunted down. Both the lion and the jogger were single mothers, and two foundations were formed and advertised, one for the cubs and one for the kids. The one for the cubs drew ten times the donations that the one for the kids did. That's rational. A lion cub in captivity is cute and interesting. An extra person in the United States is of some little value to most of us, and by your own arguments, an extra person in Africa or Asia or the Middle East or Latin America is of *negative* value to us in America; that extra person risks making those that I, Bert, most care about, my family and friends and the society that sustains them, a little bit worse off. An ethical theory that says that merely by copulating and then delivering a child, two people we don't know and share no political culture with create a moral obligation on us and everyone else richer than them to support that child—that ethical theory is untenable. It is just an invitation to Malthusian destruction, in which the most self-indulgent, the most thoughtless, the most careless, the most ignorant, create an exponentially increasing burden on everyone else. American policy should be triage, not charity to the masses of people on the planet who through their government's incompetence and corruption and their own reproductive proclivities are collectively what economists call 'utility monsters,' who would absorb all of the goods and services the world could provide and more if we were willing to try to provide them. And finally, the IPCC report is a fraud!"

"What do you mean a 'fraud,' Bert? Are you some sort of conspiracy crank?"

"Nothing of the sort, but the report is obviously guided by political correctness more than science. We have a whole turgid volume on the physical science but not a word about population growth. We have a whole turgid volume on mitigation of climate change, but the long 'Executive Summary' says *nothing* about birth control. You just have to look at their charts to see that carbon dioxide increases since 1970 closely track population growth. I read a report by a guy at the UN who figures that *half* of the coming increase in carbon dioxide emissions will be due to population growth. But talking about world population control offends the Africans, offends the Catholic Church, offends various liberal nitwits, and so the UN report won't touch it. That's really *political* science."

"But Bert," I would object, "as you say, the greatest rates of population growth are in the poorest countries—but they generate the least greenhouse gas. Africa, for example, has about 14 percent of the world population, but only produces about 8 percent of the greenhouse gas emissions. Most of the greenhouse gas emissions are in wealthy and almost-wealthy countries, and the biggest increases are in the United States and China."

And Bert would say: "You're forgetting deforestation. I read somewhere that deforestation accounts for about 17 percent of the increase in carbon dioxide in the atmosphere. Where are they deforesting? In Africa, Indonesia, Southeast Asia, Latin America, that's where. Not here. We're *adding* forests and have been for a long time. More people in poor countries, fewer forests. Its simple, and the IPCC would not face up to it; I bet the UN would have fired them if they had. And when they cook in poor places, they use stoves that send all sorts of carbon particles into the atmosphere, like a billion barbecues, and when that stuff lands on ice it absorbs heat and melts glaciers."

"But Bert, a famous political economist, the late Julian Simon, showed pretty conclusively that the general welfare increases with population, not decreases; and if you want women to have fewer children, they first need food and clinics and roads and jobs and an industrial base big enough to provide some kind of social welfare system. And the Millennium Project in Africa has shown that the standard of living in rural villages can be raised with simple, coordinated, inexpensive strategies."

"Hogwash. I read some of Simon. What he argues is that if population growth is slow enough, you get these benefits. I think the idea is that population growth has to be slow enough that the industrial base, communications, roads, water supplies, electricity, education— the whole thing—can keep up with it. Otherwise people breed until everything is eaten but the dirt. Simon's arguments just don't apply in Africa. The population isn't growing bit by bit; it's exploding. Look at Uganda: the population was about 5 million back in 1950, it's about 25 million today, and the projections say it will be 55 million in another fifteen years. And Madagascar! Too ugly to talk about. I saw on TV the other night a bunch of Westerners trying to save the Madagascar forests for the animals by getting the Madagascar government to declare them wildlife sanctuaries. In the army we called that sort of thing pissing into the wind. Besides, according to the IPCC, if we wait a hundred years while we *hope* that the poor countries stabilize, stop deforesting, and don't create more industrial pollution, we'll all be breathing carbon dioxide. And as for the Millennium Project, the guy who runs it, Jeffrey Sachs, seems a sincere, incredibly persuasive, energetic guy; and I'll bet he makes some villages richer, and I even bet that if he had the resources and the African governments got out of the way, he could make a lot of poor African people richer. But in his village projects he just ignores population control, so far as I can see. And he probably ignores it because the villagers wouldn't accept it, or their governments wouldn't, or his donors wouldn't, or all three. So that's going to make the carbon dioxide problem a bit worse, not better. Unless population is controlled, the choice is between wealth and climate. I think Sachs would choose wealth for Africa—and that's my choice too, for America. Sachs, I admit, came around. He has another book with some ideas about limiting population growth, but I don't see that it's making a big impact."

"But Bert, if people grow food more efficiently, they won't need to burn down forests for new agricultural land or cut down forests to sell wood."

"Dream on. One or two generations, fifteen to thirty years, and the population growth will overtake the increase in food production."

"But Bert, we can reduce how much carbon we put in the air in America. There's not much we can do about controlling world population growth. Respected popular scientific journals like the *New Scientist* say it's not worth discussing."

"Nonsense. How much foreign aid does the United States give out? Something like $23 billion to something like a hundred and fifty countries, not counting contributions to international food and health organizations, and not counting private donations, which are huge. So here's a proposal: not a dime in any kind of aid—not military aid, not food aid, not humanitarian aid, not drug-control aid, not anything—to any country that is not putting an effective population-control program in place, including advertising, free contraception, distribution of birth control implants and devices, people on the ground in rural areas, the works. You say this climate stuff is the biggest problem we face; OK, population growth is half the problem, so that's where you should want to put half the money. Plus there is all the private money, charities and such, Mr. Clinton's African AIDS guilt trip over what he let happen in Rwanda. Make a law restricting charitable transfers unless the receiving countries are seriously getting a handle on population growth. We could do it tomorrow if we wanted to. Some countries won't take the deal, but a lot of them will. And that's just a start. The U.S. is still the biggest market in the world. Countries want to sell us stuff. So we could put on an import duty in proportion to the exporting country's population-growth rate or, better, related to how seriously a country is trying to control population growth. Nobody can make me believe they are serious about climate control unless they endorse and follow through on a serious population-control program. And I don't think we are helpless to do anything about it. We Americans, collectively, and the Europeans and Japanese as well, we just don't want to."

"So you want forced abortion like in China, Bert?"

"I thought you were some kind of logician. How does it follow from requiring population-control methods that we should welcome or tolerate forced abortion anywhere? It doesn't. Did I list forced abortion in my account of population-control programs? I did not. So let's have an honest conversation here."

"OK, Bert, population is a problem, but it's still in our interest to reduce our own national carbon output in every feasible way we can. Global warming may do less damage to North America than to many other places, but on the whole it doesn't look like a good thing, even for us. Billions of people are not going to die quietly of hunger and thirst; they will go to war—in Africa, Latin America, central Asia, a lot of places we import oil from or plan to. And war means disruption of discovery, extraction, and transport of oil. What are you going to run your Buick on? Even if you don't care what happens to people in Asia and Africa and Latin America and the Arctic, global warming is going to disrupt oil supplies to make gasoline for your Buick."

"Who said I don't care about those people? I just care much *more* about us—me, the family, America, even you liberals."

"OK, I take that back. But here's the point. You can afford five-buck-a-gallon gasoline, but twenty dollars a gallon? And even if you can afford it, the people who grow food and the people who transport it and almost all the people in the nation, including your children and grandchildren, cannot. And winter and summer will still be around. They won't be able to heat or cool their houses either. We're talking major economic and social disruption in America. But that's just the beginning. If we keep importing and burning oil, a lot of people you don't like will own America. China holds over a trillion or so dollars in U.S. debt, and the day it floats the yuan—and eventually it will—the dollar will plummet and America will see inflation like never before. Our big financial institutions have essentially become subprime borrowers from Arab and Asian nations, promising high interest rates.

The Saud family—the plump thug group that runs Saudi Arabia, treats women like possessions, and subsidizes Islamic fundamentalism in the country that produced fifteen of the nineteen 9/11 hijackers—controls our economy indirectly. All they have to do is reduce pumping oil, and American screams. The same policies that would reduce carbon emissions in America would reduce or eliminate our dependence on such nations."

Bert might draw two lists on a napkin, like this:

Conservative Technologies	Liberal Technologies
Hydrogen fuel cells	Tiny cars
Mirrors in space to deflect sunlight	Hybrid engines
Nuclear power plants	Tiny apartments
Sulfur in space	Public transit
Iron in the ocean	Bicycles
Fence on the U.S./Mexican border	Windmills
Corn ethanol	Solar cells
Houses	Cellulose ethanol
Buicks	Shivering
Population control	Prayer

Bert would say: "The second President Bush promised a hydrogen economy, and General Motors has already developed a car, maybe a Buick, that runs on a fuel cell. Nuclear power plants generate electricity without generating much carbon dioxide, and France, which generates most of its electricity from nuclear power, has already reduced its carbon emissions."

(Aside: Bert loved France and the French. He fought through France and Germany in World War II, and after the German surrender he was sent to Biarritz for R & R. Ill and feeble, he made a sentimental return there many years later. My wife accompanied him and bribed the hotel concierge to send him up flowers and a bottle of wine with a note in good French thanking him for helping to free France, as though they remembered him. Bert was conned. Only time I know of he turned teary.)

"Many cars already on the road can use fuel that is 15 percent ethanol, and General Motors has promised cars that run on 85 percent ethanol. We can grow corn like no other country. There are serious proposals for lofting small mirrors into the stratosphere to reflect sunlight back into space, and there are big experiments seeding the ocean with iron to increase the growth of algae that absorb carbon dioxide. Some atmospheric physicists, including a recent Nobel Prize winner, have suggested that shooting sulfur into the upper atmosphere, where it is turned into sulfur dioxide, which absorbs sunlight, would permit us literally to tune the temperature of the globe, higher or lower as needed. Windmills and other 'clean' energy can at best generate only a small fraction of the energy we need, and we would need a solar-cell blanket the size of Nevada, or at least New Jersey, to generate the quantity of electricity we need."

And I would tell Bert: "Your conservative technologies either won't do the job or are positively dangerous. Putting mirrors in orbit that reflect sunlight back into space could, according to computer simulations, succeed in lowering Earth's temperature. Carbon dioxide and other greenhouse gases would still build up, but the mirrors could in principle create a 'global dimming' of sunlight reaching Earth sufficient to counteract their effect. But what would that much reduction of sunlight do to rainfall and vegetation on Earth? We don't know. The Nobel laureate who most recently considered shooting sulfur into the atmosphere did not recommend it, because the sulfur would produce other reactions—reducing ozone, for example. He merely called for research on other chemicals that could do the job harmlessly, but he had none to offer. As for iron in oceans—and other geoengineering schemes as well, let me tell you a cautionary story about the chemical engineer who eliminated engine knock and made refrigerators safe. [I will skip the cautionary story—if you've read this far, you know it.]

"Nuclear generating plants may be a good idea if we can figure out what to do with the radioactive waste, but electricity isn't going to move your Buick very far unless you want to recharge every few miles or unless there is some real breakthrough in battery technology, and electricity won't fly planes at all. Hydrogen fuel cells are a fantasy solution. Bush and GM advertised the hydrogen solution to get people to stop thinking about cars that get real fuel economy *now*, cars that Toyota and Honda, but not GM, build. You can get hydrogen in two ways: by electrolyzing water or by 'reforming' petroleum distillates. Using electrolysis, hydrogen energy costs about three times the price of gasoline, and by 'reforming' you don't get away from generating carbon dioxide and depending on the Saudis. Plus hydrogen is a gas that is hard to confine, and when it does get out it reacts explosively with oxygen. There is no safe way to transport hydrogen around the country and no safe, feasible way to store it in automobiles that can travel any distance.

"So too with ethanol, Bert. Planes will fly on ethanol but not as efficiently as on kerosene, and there are some problems—for example, ethanol likes to combine with 5 percent water by volume, so as a fuel it has to be kept away from water to work well. Ethanol has been proposed as a fuel for cars since early in the twentieth century, but corn is just about the worst way to get ethanol. After the energy inputs required to grow corn and to turn it into ethanol, corn ethanol gives a small net energy gain and doesn't do much to reduce carbon dioxide emissions. I think General Motors jumped on ethanol because they didn't have good hybrid engine technology and didn't want to invest in developing it, and of course corn farmers jumped on the idea for the obvious reason. But making ethanol from corn by present methods generates a lot of carbon dioxide and requires a lot of energy input, usually from fossil fuels. Sugarcane and sugar beets are better sources—a sugar beet root is about 15 to 20 percent sucrose, and sucrose, as any undergraduate chemistry student will tell you, makes fine alcohol. We can, and do, grow sugar beets in America, but we have an absurd fifty-four-cent-a-gallon tax on imported ethanol, chiefly to fatten the incomes of American farmers. The sugar and corn lobby says the tax is to make up for the difference between what Brazil, for example, pays its workers compared to what American workers are paid. Such hogwash! Sugarcane is harvested in America by machines, and, for comparison, American mechanized corn farming produces corn more cheaply than anywhere else in the world. If the sugar and corn lobby wants us to introduce an excise tax on *all* imported goods in proportion to how much foreign laborers that produce those goods are underpaid by American standards, I'm all for it. The farm lobbyists haven't made the suggestion, and you can be sure they won't. Maybe the unions should. Anyway, the best deal is to make ethanol from plants with a lot of cellulose. Those kinds of plants—like the switchgrass George Bush once referred to—don't yield as much ethanol by weight as sugar beets do, but they are plentiful, and there is essentially no carbon dioxide byproduct in making the ethanol.

"So, Bert, you could someday, if General Motors ever made the right engines, and if ethanol distribution systems ever got built, and if Congress ever had the guts to defy the farm lobby, drive a Buick on ethanol. You wouldn't be doing the poor of the world any favor: land that grows feed stocks for ethanol doesn't grow food. In the meantime, you should drive a hybrid. Toyota and Honda make good ones."

Bert would say that he drives Buicks, and there aren't any hybrid Buicks. And as for those of the world who would be too poor to buy corn, or whatever, if we made ethanol for cars, we have no obligation to provide them cheap food at the cost of our transportation.

I would tell him: "There were and there are hybrid Buicks. Just about the first hybrid electric-gasoline-engine automobile ever made was a rebuilt Buick Skylark, back in 1974. Now, finally, GM has a whole line of hybrid-engine cars. But a hybrid Buick? Yes, GM sells one—in China. But the GM hybrids aren't really up to snuff yet. A Saturn hybrid gets thirty-two miles a gallon; your daughter's hybrid Honda Civic, about the same size and price, gets forty-eight miles a gallon, really does. And if Americans, who mostly commute to work alone, made and drove cars like the old Honda Insight, they could get sixty-four miles a gallon in a two-seater. America isn't really trying. Besides, the way GM is going, there may not be any new Buicks to buy."

"Clark, Obama will save GM and Buicks, so I don't worry. Why he fooled with Chrysler, I don't know. He said it was because Americans could make affordable fuel-efficient cars if foreigners could, but then he sold Chrysler to Fiat, and I think they are going to make small cars in Mexico. Since when did Italians designing cars to be made in Mexico become Americans?"

Bert would continue: "OK, suppose it's in Americans' interest to reduce carbon emissions. There are two ways for Americans to do that. Americans can consume less or there can be fewer Americans. Every immigrant to America soon adopts an American level of consumption, and if they don't, their kids do. Since the legal American population is already declining, the way to reduce the number of Americans is to stop immigration, legal and illegal. That's what the fence on the border is about. Good fences make good neighbors. And if necessary, bring the army back from South Korea—time the South Koreans took care of themselves anyway, and our soldiers are more like hostages than a trip wire—and put the 27,000 soldiers on the border."

"But," I would say, "fences don't work—people go under or over or around them. And if they work, they will kill people, leave them dead of thirst in the desert."

"Piffle," Bert would reply (or something stronger, although I never heard Bert swear), "the *New York Times* crowd and the Mexican American claque are opposed to fences exactly because they *do* work. If we put a real fence all along the border, there would be no place to go around, so border crashers would have to go over or under, slowing them down. It's a long hot desert to carry a ladder across. When they slow down, they get caught and sent back. When enough of them get caught and sent back, they will stop coming. Besides, people die every day all over the world whose lives would be saved if we brought them to America. Mexicans have no more claim to come here than Ethiopians or Chinese or . . . you name them. We aren't obliged to let them in, and the fact that they are close enough that some of them die trying doesn't create that obligation either. If we did let everyone in who is willing to risk death to get here, we would have more of the global warming you're worried about until we wouldn't be America anymore. We would equilibrate to overpopulated misery."

"What evidence do you have," I would counter, "that the 'New York Times* crowd,' and whoever else, is opposed to fences because they do or will work?"

"Because," Bert would argue, "*New York Times* editorials argued over and over that rather than putting up a fence we should punish employers who hire illegals, but as soon as Arizona passed a law that really would punish employers for hiring illegals, the *Times* denounced it. The *Times* is just plain opposed to stopping border crashing but doesn't have the guts to say so. If Bill Keller and company thought a fence wouldn't work, they wouldn't bother to denounce it."

"Bert, we're all in this together, everybody on the planet."

"Maybe so, but we aren't getting out of it together."

8
Sacrifice of the Lawn

Philosophers love moral puzzles: a trolley is running down a track and will kill six people unless you throw a switch to divert it to another track, where it will kill five people. Have you killed five people or saved six or both, and are you culpable for killing the five? A fat man is stuck, head out, in the only exit from a cave rapidly filling with water. Five people are in the cave, and they have a stick of dynamite the perfect size to disintegrate the fat man. Should they do so, morally? The puzzles, if that is what they are, are meant to challenge naïve moral rules and to raise questions about very general moral principles. Do the numbers killed count? Is causing a death as heinous as declining to save a life? Which of our choices are morally up to us and which are constrained by moral duties and obligations?

Novelists and Hollywood screenwriters fashion more realistic and more vivid moral explorations—*Sophie's Choice, The Cider House Rules*—but real life has created the moral problems with the greatest consequences, and one is upon us. Twenty percent of the world's population is malnourished, and thickets of people are newly hungry. They are, quite literally, starving in Africa . . . and parts of Asia and Haiti and elsewhere. Who should get the blame, who should feed the hungry, and how?

The world food crisis of 2007–2008 cast biofuel production in a harsh light; it was denounced almost everywhere as a cause of world hunger, concomitant riots, and even the overthrow of a government. Was it? Biofuel production took up a tiny fraction of world agricultural production that might otherwise have been used for food. Despite a lot of blogging and posturing to the contrary, biofuel production in 2007–2008 does not appear to have substantially lowered world food production from the levels of previous years. Corn exports from the United States increased rather than decreased, even as corn was diverted from food to ethanol. According to the United States Department of Agriculture, corn exports from the United States increased in 2007 by nearly 400 million bushels. One might think that happened because corn acreage displaced other grains, such as wheat. But wheat exports increased as well, and soybean exports were down only slightly. Rice exports in 2007 decreased only slightly from the previous year.

One thing that did change was world food reserves, which dwindled systematically over several years and began dwindling well before the rush to biofuels. World food prices had been increasing steadily for several years. The dramatic jump in 2007–2008 is most plausibly attributed to reduced supplies from critical regions, such as Australia; increasing demand; and dramatically increased fuel costs, which affect everything associated with food production and export: fertilizer and the fuel for tractors, trains, ships, trucks. World food prices track oil prices. Destroying OPEC would be the best way to keep world food prices down.

Biofuels haven't done much damage to food supplies yet. But they may eventually. So the question is: should biofuel production be stopped and the land returned to food production? Is that a moral requirement?

In advocating moral responsibility for a change of behavior or policy as a remedy for a problem, there are at least four kinds of questions. Is the policy important for other pressing problems? How much is the policy in question a cause of the problem? Are there possible alternative changes in other policies or behaviors that could ameliorate the problem and that are at least equally obligatory? Do those whose behavior would have to change to remedy the problem have a duty to do so?

The moral case for substituting biofuels for petroleum is diverse: properly produced, biofuels can substantially reduce greenhouse gases (corn ethanol is not a proper way; sugar beets and cellulose alcohol are better); substitution of biofuels by Europe and the United States will reduce petroleum prices worldwide, which will aid less developed countries that are not oil exporters, countries for which increased fuel prices are ruinous; reducing Saudi profits will help to undercut a vile nation that uses its wealth to campaign across the world against human pleasure and rationality. The case against biofuels is simpler: they reduce world food supplies.

World food shortages have a lot of causes, but the principal causes are four: oil, weather, population, and thuggery. The food production of Africa has been stunted by drought and the government of Robert Mugabe in Zimbabwe. African heads of state were quiet or cheered as Mugabe decimated that nation's agricultural output. The main shortage in 2007–2008 was in rice; the shortage was principally caused by a multiyear drought in Australia, but also by the thugs running Burma and by national hoarding. Food substitution is not always feasible. Some years ago, when a famine struck a rice-eating region of India, the citizens burned a relief train full of grain because it carried wheat, not rice. No reason to think corn would have received a better welcome.

Wherever they can afford to, people convert in huge numbers to eating meat, and the calorie loss in the conversion of grain to meat is enormous. The green revolution in India in the past decades increased the average calorie consumption in that country, but it also dramatically increased the nation's meat consumption. Something similar is happening now in China. World population is (at this writing) 6.3 billion people, more or less—probably more—and growing; an increasing proportion of it can afford to eat more and does so. Many of the rest eat less.

Thugs and weather and, I suppose, the oil cartel will always be with us. Can we make more food? The United States has vast farmlands sequestered from growing crops, but even so an obvious question is why policy and responsibility should focus on the use of farmland for biofuels rather than on other ways of increasing the world food supply. The fastest-growing agricultural land use in Australia in recent years has been in vineyards, and huge areas of California, France, Spain, Italy, Latin America, and North Africa are devoted to grape production. Worldwide, roughly 20 million acres are devoted to growing grapes, mostly for wine. Some of the soil currently used for grapes could be used to grow more nutritious crops. If there is a moral imperative to use agriculture to feed people, then surely other crops take precedence over the making of wine. So there is one solution: rip out some of the vineyards. There are others.

In 1914, Carol Aronovici, the general secretary of the Philadelphia Suburban Planning Association, wrote in the *Annals of the American Academy of Political and Social Science:*

> The city has, humanly speaking, proven to be a failure. Congestion of population and concentration of industrial activity have been over-capitalized and no contingent means have been provided to meet the needs for a normal human development and efficient industrial growth of our cities. These two important factors are now pointing the way toward a hopeful solution of ultra-urbanization of all human activities. The decentralization of human habitation first found expression in the splendid development of our metropolitan suburbs . . .
>
> It is through this exodus that we hope to solve a considerable share of our housing problems, improve living conditions and create a closer cooperation and deeper sympathy between the worker and his work.[1]

Times change. Nowadays, urban theorists tend to say that the American suburbs are an ecological disaster: pavement and malls, houses too large, huge transportation costs. But the American suburbs are a significant unused potential source of world food. My place is not atypical: half an acre of unused pasture, sufficient with my neighbor to keep a miniature milk cow or a goat, a pig, chickens. Yards less steep could grow corn or soybeans, or . . . The lawns of America could, I reckon, easily replace the food lost to biofuels. Marxist sociologists argued long ago that the whole point of the suburban lawn is to show

1. Carol Aronovici, "Housing and the Housing Problem," *Annals of the American Academy of Political and Social Science* 51 (1914): 1–7.

the world that its owner can afford to waste perfectly good land that could be used profitably to grow a garden or feed a cow or both. The laws of suburbs suggest that the Marxists were right about that. In much of suburban America you can keep a dog as big as a pony, but you cannot keep a pony. You can legally keep as many parrots as you like but not a single chicken, not a single guinea fowl, turkey, rabbit, pig, goat, sheep, or cow. In places where the suburbs become a bit countrified, the bigotry against farming does not lessen. Land in the Chicago suburbs is not cheap. If you want to keep a horse in any of the counties bordering Cook County, their laws require that you own at least five acres. But should you want to keep a chicken or a goat, well, not less than fifteen acres will do. The discrimination cannot be aesthetic: compare the clamor of four dogs barking to the clucking of a hen.

So a solution to a real world food shortage, should it come, is to tear out some of the vineyards, repeal the laws that make the suburbs agricultural castrati, plow the golf courses. In the meantime, grow more, and more efficient, biofuels. The alternative means to feed a population of 9 billion or so would seem to be *de minimus* living for almost all: eat locally, be vegetarian, make compost, and all that. We hear already that obesity is no longer just a health problem—it's a moral problem. Those who eat too much keep others from eating enough, the logical inverse of my mother's illogical injunction to eat everything offered because of the starving Armenians. The economics of that argument seems a bit shaky, but even I, a lackluster slob if ever there was one, look with some disapproval on wanton waste—people idling their motors to no point or gadding about the suburbs in a Mega-Hyper-Battleship Galactica SUV fitted out for crossing the Sahara. Eating meat is a pleasure, not a necessity, for human life or health, and surely it is morally wrong to kill merely for pleasure. The late Robert Nozick made this point with one of those philosophical stories: would you consider immoral someone who went about clobbering cows with a baseball bat because he really enjoyed the feel and sound of a bat striking a living bovine skull? I would. Why then is killing the cow for the pleasure of eating meat any better? Hard to say. In Nozick's story, is it the sensation that gives pleasure?—in which case inanimate surrogates could be used—or is it the knowledge that a living cow's skull is being smashed? If that, then the story is not entirely on a footing with killing the cow for meat, which requires no pleasure in the killing. Quite the contrary: Americans and Western Europeans are squeamish, which is why we have our killing done far away and out of sight. Of course, we carnivores could hope for a credible surrogate, with the taste and smell and look of prime rib.

In the meantime, with misallocations of food distribution but no real world food shortage, vegetarianism marches hand in hand with a movement for "eating locally." Eat tomatoes and corn in the summer, potatoes and turnips in the winter; compost your leaves and leftovers. I oppose these efforts, as I oppose all forms of puritanism. It is true enough that the vegetables from your garden or your local farmer are fresher and probably taste better than those from far away; but unless you live in blessed climes, those local delicacies are not available eight months of the year, and the wines of Montana and Lake Erie will never compare with those of France. Denying the pleasures produced from afar is no more effective in reducing world hunger than saving the starving Armenians by cleaning your plate. And as for composting, that supremely inefficient process to turn your garbage into fertilizer, a goat will take care of the leaves; and for the rest there is a faster way, with an edible byproduct: get a pig.

III

Science?

9

John Henry at NASA

I was once a spy for NASA, and I once set NASA against the steam drill, figuratively, and the steam drill won.

In 1997 NASA's dramatic Pathfinder mission landed a mobile robot on Mars for the first time. Pathfinder was an engineering triumph, but the instrumentation carried on the mobile robot was comparatively elementary. Pathfinder (like subsequent Mars surface rovers) was operated entirely from Earth. People told the robot where to go, when to stay still, and what instruments to record with, and Pathfinder sent back to Earth all of the data it recorded. Pathfinder was dumb. Sending back data from Mars to Earth took some energy, which Pathfinder did not have a lot of, and receiving the weak signal required the time of big, busy antennas on Earth. Nothing could be sent or received if Pathfinder was out of the line of sight of the Earth antennas. For some of NASA's scientists and engineers, Pathfinder's limitations posed an opportunity: why not build a mobile robot that could guide itself, choose where and when to record data, analyze the data itself, and then send the scientifically relevant summaries and conclusions back to Earth? (There is no better form of data compression than scientifically relevant conclusions: everyone sensible knows that Earth revolves around the sun, but only weird scholars actually know the data Copernicus used in arguing for that conclusion.) Why not, in other words, send an artificially intelligent robot to explore the surface of

Mars? The need for antenna time would be dramatically reduced, the robot could work out of the line of sight of Earth, and it would not need to store all of its data, only the data it was using in assessing Martian conditions at any moment. A robot like that would not be as good as, say, having an expert geologist on Mars—more like having an understudy geologist on Mars. So began what came to be known at NASA Ames Research Center as the Graduate Student on Mars project.

NASA Ames had a history with artificial intelligence and space exploration—an unhappy history. Deep Space I was launched in October 1998 to explore asteroids and a comet—without landing on any of them. Part of the mission was to test new technology, and part of the new technology was the Remote Agent software. Remote Agent was a project to develop and test a kind of primitive HAL, of *2001* fame: a computer program that would guide the spacecraft, plan when its trajectory should be changed and how it should be changed, direct the changes at the appropriate time, guide the direction of the optical instruments so they were focused on the appropriate asteroid targets, and so on. While the Deep Space I mission was overall a great scientific success, nearly everything that could go wrong with Remote Agent, short of crashing into the sun, did go wrong.

Remote Agent used a variety of artificial-intelligence techniques. For example, part of Remote Agent was built around methods of "nonmonotonic reasoning," which were at the cutting edge of artificial intelligence at the time. The basic idea is this. Suppose the robot thinks: "All birds can fly." Suppose it also thinks: "Ostriches do not fly." Using deductive logic, the kind of reasoning that Aristotle first described in the fourth century B.C. (and not improved on until the nineteenth century by two philosophical mathematicians, George Boole and Gottlob Frege), the robot can infer: "Ostriches are not birds." Now suppose the robot learns two new facts: "Tweety is a bird" and "Tweety is an ostrich." What is the robot to think? The robot's beliefs entail that Tweety cannot fly, which is correct; but they also entail that there exists a bird that cannot fly, which is also correct but contradicts one of the robot's other beliefs, that *all* birds can fly. In classical logic, any single contradiction implies *everything*. An intelligent robot should be able to draw logical conclusions, and if so, the robot could be deducing consequences forever, and most of them would be absurd, for example, "Round squares are hungry." Nonmonotonic logicians, chiefly people from philosophy and computer science, found a variety of ways to control such situations with formal, computable systems of reasoning, which the robot could use to avoid deriving absurdities and could recognize that in the circumstances "All birds can fly" becomes a kind of rule of thumb, with exceptions. A computerized control system needs this kind of reasoning ability since it may at a certain time believe the state of the spacecraft to be thus and so, but then it receives information from sensors that say otherwise. The new information may be correct or may be due to a faulty sensor, and so on. Classical deductive logic won't do, not even for a mini-HAL.

Remote Agent was originally scheduled to be run and tested in flight on the Deep Space I spacecraft over three weeks. The test period had hardly started when Remote Agent began behaving badly and had to be shut down. Correcting the software on Earth and sending it up to the spacecraft took a week. Things immediately went wrong again; Remote Agent had the wrong asteroid targets. The next day Remote Agent forgot to turn off rocket thrusters—part of the necessary software was missing. Humans took control and decided to skip further testing of that part of Remote Agent. (Whew!) The three-week test was now reduced to six hours. That went OK. Except that the program somehow lost values from a critical system monitor.

In their final report the Remote Agent team concluded, with upper lips entirely stiff, that "by executing the two flight scenarios, RAX achieved 100% of its validation objectives."[1] NASA administrators seem to have had a different conclusion, and I know of no further attempts to control spacecraft with mini-HALs.

So the very idea of a robotic, self-controlled scientist on Mars was

---

1. Douglas E. Bernard, Edward B. Gamble, Jr., Nicolas F. Rouquette, Ben Smith, and Yu-Wen Tung, *Remote Agent Experiment DSI Validation Report* (Jet Propulsion Laboratory, 2000), p. 21.

not going to be well received around NASA. Nor was it going to be well received by most scientists. Scientists want the data, all of it, even if there is too much for them to sift through, let alone analyze. Making scientific explanations is *their* job, their craft, and they were not eager to see any part of it handed off to some computer with wheels. But the idea found a sponsor at NASA in Ken Ford, a computer scientist who worked in artificial intelligence and had become associate director at Ames. Ford made some money available for initial research and, for good reasons, gave principal control of the project to planetary scientists, not to computer scientists.

Ford's good reason was that he knew computer scientists. They have their passions, their hobbyhorses, and what they do is esoteric. Ask your average programmer to write code, say, to simulate a dog barking, and what you will get is a simulated dog that looks like a pig and doesn't bark, along with a new, general computer operating system that doesn't quite work but which the programmer will tell you, in obscure terms, is marvelously superior to Windows. So Ford put the money in the hands of planetary scientists, who at least knew what the point was. But he was concerned that the computer scientists, who would inevitably get most of the money, might pull wool—they had a whole flock of the stuff—over the planetary scientists' eyes. So he asked me to take part in the Graduate Student on Mars meetings and let him know about any wool. I was, as it were, neither sheep nor wolf. That worked for a while, but eventually a full moon rose and I became a wolf.

The first meeting of the project was attended by several young computer scientists from around the Bay area, mostly postdocs or graduate students of senior computer scientists. Planetary scientists led the meeting and explained the project goals, giving an illustration of work by a computer science student, in collaboration with a NASA geologist, that identified rock and gravel slides on Mars. The planetary scientists wanted to know the mineral composition of Martian rocks and soils, and they wanted to be able to identify geological formations that might have been formed in the past by water flowing on the surface of the planet. Water means life, maybe. The planetary scientists emphasized that future mobile robots sent to Mars would, like Pathfinder, have rather small (compared to a desktop computer) memory storage and relatively limited computing power. Supercomputers were out.

The computer scientists zoomed in on what interested them: programming languages. There are lots of programming languages— from Logo and BASIC to FORTRAN to Pascal, C, C++ to . . . It's endless (although I don't know what became of A and B), and new ones are invented almost every day, while old ones fall into disuse. Java was then the new computer programming language du jour, so the computer scientists argued that what was needed was research on how to make Java have a small "footprint." Computer programs, when compiled into a computer's memory, take up varying amounts of space depending on the structure of the program and on the programming language; Java is big. So the planetary scientists should invest their money in the study of programming languages. The NASA planetary scientists, who were utterly clueless, left the room muttering that maybe they should put all of their money into research on Java.

At the time, the NASA guys had no idea how to do the main thing, even if they had a supercomputer on Mars: have the computer take data from instruments on a mobile robot and turn it into information of scientific interest, then have the computer use its instruments to decide which distant targets were worthy of investigation. So they could put their money into figuring out how to do the thing at all, or they could spend it subsidizing work on programming languages. Eventually, with a little prodding, the planetary scientists worked through the options and decided to figure out how to solve the main problems. Programming languages would have to wait. (Eventually, and typically, NASA decided to develop its own programming language for robotic software.)

Ford was sound, or lucky, in his choice of the main planetary scientist for the project, Ted Roush. Roush is a handsome, quiet, lean man who often bicycles to work and back through California traffic, and he is a real-McCoy expert on how to figure out what something is made of from the structure of the light reflected from it. More than that, he was genuinely open-minded and curious as to how well computer programs could do that part of his scientific job.

Isaac Newton showed that if sunlight is passed through a glass prism and onto a white surface, the light is spread into bands of different colors—a spectrum. Ever since the nineteenth century, the bands have been labeled with their frequencies, which indicate different energy levels of the "components" of ordinary sunlight. Depending on their composition, surfaces struck by sunlight will absorb light at particular frequencies, turning it into internal energy in the particles of the surface, and send back the rest. So if sunlight is reflected from a material surface that isn't white and then the light is passed through a prism, components of light at some of the frequencies will be missing or diminished, and the result will be an absorption pattern. Now suppose one knew, say from laboratory experiments, the characteristic absorption patterns of various minerals of interest—for example, minerals that are typically deposited from water. Since different minerals absorb different frequencies of light, by looking at the spectrum of light reflected from a rock or soil sample and comparing it with the laboratory spectra, the mineral composition of the sample could be determined.

Sounds straightforward, but it's not, really. The first problem is that there are way too many minerals—*Dana's Mineralogy*, a standard reference book, lists more than 3,000 of them. And the minerals can occur together in rock and soil samples. If, say, up to four minerals occurred together in a sample, that would come to about 3 thousand trillion different possible combinations of minerals, give or take. We have about 135 laboratory spectra for "pure" minerals, and a few hundred more for mixtures of minerals. And then there is the sun. The energy of sunlight is distributed over an infinity of frequencies—our eyes are sensitive to a relatively small frequency range, but one in which a lot of the sun's radiant energy is packed. The problem is that the energy distribution of light reaching the surface of Mars is different from that on Earth and different at different times of day. So the spectrum of light reflected off a soil or rock sample has to be corrected to match a known sample, usually a laboratory sample. The standard way is to have a white reflecting surface next to the target, both in the lab and in the field, and to adjust the intensity of light at each frequency by comparing it with the intensity of light at that frequency reflected from the white surface. That creates a problem for a robot that wants to use light reflected from distant rocks or surfaces to decide whether the rock or surface is worth investigating more closely. What is the poor thing to do—run up to the rock from a distance, put down a white card, then run back to where it started and measure the spectrum? Even Pathfinder wasn't that dumb. Pathfinder had aboard thin samples of a few minerals, which were exposed to the Martian sunlight just as the rock and soil samples were whose spectra were taken. So the samples could be compared with known minerals in almost the same place at the same time. Another way, when the instrument—the spectrometer—isn't too far from the target, is to have a white surface near the front of the spectrometer, so that the instrument can compare the not-too-distant target and the much closer white surface. Doesn't work so well for light reflected from planets or light reflected from space garbage, but it might work on Mars.

I got interested in how a computer could identify important mineral components in an unknown sample, and with a graduate student, Joseph Ramsey, I went to work on the problem. Roush suggested we focus on carbonates, minerals that have a CO_3^{-2} component. There are about twenty different known carbonate minerals, but only two, calcite and dolomite, are at all common on Earth. Calcite is calcium carbonate. Dolomite is a double carbonate with calcium and magnesium. No one knew whether these, or some of the other carbonates, occurred on Mars. Eventually we produced a program that was able to identify many kinds of carbonates from Earth samples, and Ramsey wrote the only doctoral dissertation ever in philosophy called "Mineral Classification from Reflectance Spectra," but that is not the point of the story.

If we were going to make a graduate-student computer for Mars, we needed a baseline; we needed to know how well a human expert would do with the same problem and with the same data that would be available to the robot. Although the spectra of reflected light had been used for more than seventy years to help identify minerals, we could find nothing at all that tested how good human experts are at the job. Absolutely nothing. Either human experts just hate to be tested, or no one had thought of it. But we needed to know. So we made up a test, and we gave it to Ted Roush, aka John Henry.

We gave Roush a computer program that would present spectra from a couple of hundred or so different mineral samples, all of known composition and about half of which contained some form of carbonate. With each spectrum there was a checkoff list of seventeen possible classes of mineral components, one of which was "carbonates"—carbonates of any kind. Roush could examine the spectra at his leisure, consult reference books, turn off the computer and go away, and come back later to the same problem, almost exactly as if he were working on data sent back to him from Mars.

Classification of samples is very much like chemical distillation. Distill off ethanol from a vat, and what you eventually get is a mixture of ethanol, water, and fusel alcohols (propanol, butanol, and other nasty stuff). If you distill long enough, you can get the water content down to 5 percent, but that's it (unless you put the liquid through some substance that absorbs water). Some of the alcohol is always left in the pot you distilled from. The same is true with recognizing carbonates—some samples with carbonates will be correctly identified as having them, but so will some samples that do not in fact have carbonates, and some carbonates will be missed. The aim is to find methods that get the most intended targets with the fewest unintended nontargets, but you can't maximize intended targets and minimize unintended nontargets at the same time: there is always a trade-off. By one trade-off, Roush, a world expert, did terribly on our test: he missed three quarters of the samples with carbonates, deciding they were not carbonates after all. By another measure, he was almost perfect: with one exception, whenever he said a sample was a carbonate, it was. The surprise was that the carbonates he identified were all calcites and dolomites—the most common carbonates on Earth, the kinds Roush had experienced in his scientific career—and he identified all of the calcites and dolomites in the test samples. Our program wasn't quite as good as he was on the dolomites and calcites, although it was pretty good, and it found about half of the other carbonates he missed.

Roush was a lousy carbonate detector, but he was an almost perfect calcite and dolomite detector, better at that than our computer program. He was genuinely expert, but not quite at what he would have thought he was expert at. Here is a moral about expertise: test it, carefully. There is an enormous low-brow psychological literature about expertise on all kinds of things, from medicine to air traffic control, which discusses how experts are like and unlike novices, how they communicate, how they part their hair and whether they lisp, but less about the essential thing: what, if anything, are they expert at, and how expert are they?

Roush was delighted with the insight into himself, positively fascinated, and he wanted to know whether he was singular. Would another geochemist, trained in reflectance spectra, have test results like this? So he hired one. She was a relatively recent Ph.D., and she went to work as part of the Graduate Student on Mars project. One of the first tasks that Roush set her was to take The Test. Some months later she had not and, when pressed, explained that The Test was programmed for a Windows machine and she could use only a MacIntosh. So we reprogrammed it for a MacIntosh. Some further months later she still had not taken The Test. When Roush pressed again, she resigned.

Over the years I have asked various experts—statisticians, psychologists, planetary scientists—to test themselves against one or another computer program. Ted Roush is the only one who ever accepted the invitation.

I did some more work on the Graduate Student on Mars project, including an experiment in the Mojave Desert, not far from where I hunted rabbits as a kid. No more—I got heat stroke and spent two days in a motel, incoherent. Reflectance spectrometry became a more or less dead issue for Mars exploration—it doesn't get beneath the surface of the rock, and Martian rocks are weathered and often coated with dust. Other methods of identifying composition have been used on Martian rovers since Pathfinder.

One day after Bush the Younger was elected, a friend at NASA called, a woman who had worked on the early space suits and had since become an unofficial NASA *kochleffl*. "Can you give me some good scientific reasons for going to the moon?" she asked. I really couldn't.

10

Total Information Awareness

Just as everyone my age knows where they were when they learned that JFK had been shot, even those a good bit younger know where they were when the planes crashed into the towers of the World Trade Center. Millions watched on television, as I did, as the second plane struck. For those in or near the towers and those who cared for them, the effect was immediate. After the shock, for everyone in America the effects were slower but enduring: two wars, thousands of our military killed, tens of thousands more maimed, the further near bankrupting of an overindebted nation, a profound change in the powers of the federal government to inquire into the lives and doings of its citizens, and—what may seem a minor consequence—a considerable change in the focus of research in computer science.

The United States Foreign Intelligence Surveillance Court was established in 1978 on the recommendation of a committee led by Frank Church, the last liberal senator from Wyoming. The FISA court, as it is known, consists of three judges who decide in secret on government requests for warrants for wiretaps or other surveillance of suspected foreign agents inside the United States. Until the (W.) Bush administration, the court granted essentially all requests, with four or five exceptions. The Bush administration under Attorney General John Ashcroft began unauthorized surveillance sometime after 2001, and in the years thereafter the court objected to more than a hundred requests for warrants. Which brings us to Rear Admiral John Poindexter and the Total Information Awareness project.

For a long while, Poindexter was about as influential as a naval officer can be. He graduated first in his class from the U.S. Naval Academy in 1958—the same year John McCain graduated from the same academy fifth from the bottom. Poindexter served in various senior commands, rose to the rank of rear admiral, and became the deputy national security advisor and then the national security advisor in the Reagan administration. And then he came close to committing treason, close enough to be convicted of felonies for conspiracy, perjury, fraud, and alteration and destruction of evidence in the Iran Contra affair, in which he and Oliver North, then a lieutenant colonel in the army and also subsequently a convicted felon, secretly arranged to sell weapons to Iran and use the proceeds to fund guerrillas in Nicaragua. Poindexter's and North's convictions were overturned on the grounds that the evidence against them was contaminated by their testimony to Congress. (Rush Limbaugh dittoheads, if any are reading, take note: North's appeal was aided by the American Civil Liberties Union.)

Poindexter was the brainy one, and his embarrassment did not interfere with getting back into government in a big way: he became an idea-and-projects guy at DARPA, the Pentagon's Defense Advanced Research Projects Agency. One of his projects was Total Information Awareness (TIA), collecting and integrating all kinds of data on just about everyone and then applying computer methods to identify those who might be a "threat." There is a lot of data available about you in one place or another: where you were born, what charities and politicians you give to, your income, your bank records, your credit card records, your jobs, your addresses, your friends, relatives, and spouses, your lawsuits, your other encounters with the legal system, your telephone calls, your Internet clicks. If the data were unified, not only could a computer track your life, a smart computer program could build a pretty accurate picture of your tastes and habits and politics and religion and character. Here was the vision: the government could know what was going on, who was doing what, all over America, and with that information, it could find those who are doing, or planning to do, or likely to plan to do bad things.

Thanks to the *New York Times,* Congress got wind of TIA and cut off funding, which meant, of course, that DARPA went on with the project in several pieces under various names. Scores of defense contractors—Beltway Bandits—rushed to help with all kinds of canned statistics, and all over the country academic computer scientists started working on intelligence and security.

The *Times* is a funny newspaper without funny pages. It has famously good investigative reporting, some really bad investigative reporting, a few columnists who know and tell about something interesting that most readers don't know (Bob Herbert, Paul Krugman, Nicholas Kristof), a few columnists who know nothing (for example, Maureen Dowd and, briefly, William Kristol), and some who are truly vacuous (Gail Collins). Most puzzling of all, some of its editorials have a remarkable incoherence (perhaps because Ms. Collins helped to write a good many of them, but there's a committee). The *Times* editorial page denounced the Total Information Awareness project for two reasons: it was a threat to civil liberties and it was a waste of money because it couldn't possibly work.

Sure it could. After 9/11 a slew of guys from one or another government office or one or another Beltway Bandit firm came to see me (and a lot of other academics) seeking pixie dust. What many of them wanted was a computerized intelligence analyst. The National Security Agency is a big, complex place (places, actually), hierarchically organized and secret. Low-level analysts receive questions and assignments—what's going on with Ali So-and-So; what's going on with such and such organization; what is the Egyptian press writing about this or that? The analysts, who are for the most part intelligent and practical, have access to all kinds of data, some of it public, some not, in the form of text, pictures, numerical tables, histories, sometimes reports of other analysts—all electronic, of course. They have pretty simple cognitive aids, effectively spreadsheets, and, depending on their roles and tasks, access to simple software for keeping track of social networks. They pass answers and summary information on to superiors, who are supposed to put two and two together. Besides analysts of various levels, of course, NSA has mathematicians, cryptologists, a social network research group, people running electronic eavesdropping, and even a few people whose job is to find new technologies outside NSA that the government might use. NSA's parking lots seem endless. But there are way too many questions to ask, way too many circumstances, too many hints to track down. NSA thinks it has a shortage of competent analysts.

An analyst has to use a lot of her knowledge and common sense, together with data from many different formats and sometimes in various languages. The guys (and almost without exception they were male) from the government who came around basically wanted a computer program that would do what the low- to medium-level analysts do, at least as well as the analysts, and faster. Many of them were sincere and naïve, and to those I talked about Turing's test.

Alan Turing helped to create modern computer science, not incidentally by pursuing the twentieth-century elaboration of an esoteric philosophical question from the seventeenth century. Turing proposed three tests for whether a computer running a program might deserve to be called intelligent: Could the computer imitate a man so well that someone communicating with a man and the computer by teletype would not be able to tell reliably which was which? Could the computer and a man each pretend to be a woman so well that one could not tell which was the man and which the computer (by teletype)? And, most seriously, could a computer be designed that would learn what an infant becoming a child becoming an adult learns?

There is no computer and no software that passes any of Turing's tests, and none are in the offing. There are programs about to appear that will try to answer just about any question posed in elementary English, but that's a long way from holding a conversation. There are, instead, an abundance of programs that solve particular kinds of problems as well as or better than humans. Typically they are problems for which lots of reasoning steps are required, or problems for which a huge number of alternatives must be searched within a well-structured space of possible answers, or problems that call for consistent application of a rule over a great many examples. These are not the kinds of problems an intelligence analyst faces. The analyst must rapidly bring to bear an abundance of knowledge about language, human behavior, social customs, and so on, most of which she cannot articulate; she must rapidly call on examples, analogies, metaphors, all woven through her experience. No computer is Miss Marple.

That does not mean that the *Times* was right. The TIA guys were looking for the wrong thing, a computer program that would simulate an analyst, rather than the right thing, a computer program that would extract the information they wanted from the data they had. The Defense Department is besieged with Beltway Bandits—contractors who claim, in effect, that they can solve Turing's problem. So a kind of test was devised. Two imaginary terrorist scenarios were developed. In accord with the story in each scenario, messages and reports were generated, most of them having nothing to do with terrorism and not involving terrorist planners. Terrorist communications were in a simple code, for example, a "birthday cake" was a bomb. All of the data documents had a date recorded, indicating the (simulated) time at which the report was made.

The first scenario was actually pretty easy to figure out from the data. A friend of mine working at a Pittsburgh software firm spent a morning with the communication logs and correctly described the plot and the plotters. The second scenario was harder. The story was as follows: The Pavlakistani head of the Needabaath Party refused to comply with UN resolutions, and the United States and its allies invaded and occupied the country, which prompted an insurgency by the Needabaathists and various Islamic radical groups, including the Ali Baba Organization. The overall plan was that the Needabaathists would attempt a coordinated guerrilla campaign. Ali Baba would broadcast and distribute anticoalition propaganda, and if nothing else worked would use "weapons of mass destruction."

The data for the second scenario consisted of more than two thousand documents recording communications or meetings or police reports. About sixty involved direct planning among nine terrorist cells. The background information—not in the data—listed names of persons in each cell and who the leaders were. The data could be assembled into entries listing who was involved in the report and what topics were discussed if the report was of a communication.

What can a computer and simple software do with this data? What the computer can do is increase the *concentration* of bad guys. Detection of a rare event or condition in a large collection of data is very much like chemistry, like producing alcohol (or gasoline): the procedure doesn't produce all of what you want and nothing else; it produces a higher concentration of what you want. Distilling produces a sample with more alcohol and less water and fusel oil. A good terrorist-identification computer program finds a subset of people with a higher percentage of terrorists. We can only hope for a computerized procedure that identifies some of the relevant groups and their members (hereafter: bad guys) while excluding most of the others (hereafter: innocents).

There are one thousand "actors"—names of simulated people involved in the communications—in the second Ali Baba scenario. The nine groups of bad guys have nine leaders. There are about sixty-six bad guys. A simple (sort of) computer procedure found a third of the bad guys, including five of the nine leaders, clustered into groups in which more than half of the members were bad guys. There's a catch, of course: a lot of innocents were identified as suspects. The concentration of bad guys was increased from about 7 percent in the original population to 37 percent in the group identified by the computer as suspects.

So what was the magic? Nothing very complicated. Suppose A reports to B and B reports to C or, alternatively, suppose A commands B and B commands C. Assume therefore that when B receives a communication from A, B is more likely to communicate with C shortly thereafter. Technical details aside, that is the only idea the program uses. When the software is changed to require that members of groups of interest share a common word or theme in their communications, few innocents are identified, and still almost a third of the bad guys are identified.

Now imagine what could be done with all of the data sought for the Total Information Awareness network. Any mediocre machine-learning guy could make a program that would help identify terrorists. With more information about people, a smaller *percentage* of those identified as potential terrorists would actually be innocents. But just as there is always water distilled off with alcohol, there will always be innocents caught in the computer's search, and even if the percentage of innocents is small, if data from all of the people in the nation are examined, a lot of innocents will be misidentified as suspicious people. But Total Information would "work," and maybe the *Times* needs some computer scientists on its editorial committee. The world has changed.

11

In and Out of Freud's Shadow

My very first semester at the University of Montana, I was a research assistant for a psychology professor. This resulted in the only psychological experiment I have ever conducted and in a certain self-understanding: I was not cut out to be an experimental psychologist.

The professor, Bob Ammons, had developed his own "personality test," which was given to every incoming freshman at the University of Montana, need it or not. The test had a scale with extremes, high and low, and my job was to interview every student who had scored in the highest 10 percent or the lowest 10 percent. To every one of them I was to give a Rorschach test and a Thematic Apperception Test (TAT), record their responses on an old (now) two-reel Wollensak tape recorder, and score the responses by formulas in the handbooks for the tests. The scoring rules made no sense to me and had little or nothing to do with the content of the responses people gave to the pictures. Everyone has seen a Rorschach card, which looks like something Jackson Pollock might have done. Several, but not all, of the TAT cards can be seen on the Web (in some places with a warning from some clinical psychologist that they ought not to be made so public—too late).

The TAT cards are all black-and-white and dramatic. I hoped they would prompt some good stories from the Ammons-extreme students, but no such luck until my very last subject walked through the door of the abandoned army barracks Ammons had given me for an office. She was tall and lean, with long black hair, dark eyes, clothing to match. She smoked. I mean she *smoldered*. What was she doing in Missoula, Montana, in 1960? She sat down to the Rorschach and I pushed the button on the Wollensak. We went on to the TAT, and all was comfortable until she reached a particular card showing a log cabin set among tall trees in a forest, with smoke tendrils coming from the chimney. She glanced at the card and then, looking me straight in the eyes, said, "A beautiful girl lives in the cabin. She is in love with a tree. It is a tall, straight, stiff tree. She loves to wrap herself around its trunk. Sometimes she kisses it. But her father is jealous of the tree. He has cut it down and now she is very, very sad." Then she left. Sweating and faint (I was eighteen, after all), I rushed to the Wollensak and pressed Rewind. Nothing happened. I pressed again, same result. The Wollensak was unplugged.

For a decade after, I put psychology pretty much out of mind. One day in August my department head came to my office and told me that in two weeks I would be teaching a course on philosophical issues in psychology. The distinguished senior faculty member scheduled to teach the course was ill, and I caught the duty. I checked: I had read one book by B. F. Skinner, one book by Clark Hull (I liked his name), and two books by Sigmund Freud. So I gave the course on Freud. Once you give a course on Freud you never get shed of it, so over the years I came to know a lot about what Freud thought and wrote; the more I read, the more apparent it became that Freud's career was a slide greased by ambition, a plunge from scientific disappointment to self-deception, perhaps to outright deceit, and certainly to silliness (read his account of why women weave in the *New Introductory Lectures on Psychoanalysis*). In 1987 John Kihlstrom, a psychologist at UC Berkeley, published an essay in *Science* contrasting Freud's unconscious with the new, scientific unconscious revealed by cognitive psychologists. He could scarcely have been more wrong, which is why, I suppose, the journal is called *Science,* not *History*. It turns out that a lot of twentieth-century psychology is footnotes to Freud, and that is not entirely a good thing.

Vienna in the 1880s was engulfed in new ideas about the mind. Two decades before, Paul Broca had described a patient, whom he called

Tan, who had lost the capacity to speak—"Tan" was his single utterance. His motor capacity was intact and he could make the usual range of sounds, but he could not produce normal speech. Upon his death, autopsy showed that an area toward the front on the left side of his brain had been damaged. Broca's description of his discovery changed how scientific people thought about the mind and brain. A long tradition divided the mind into faculties: reason, appetite, will, and so on. A generation before Broca, Francis Gall had proprosed that all sorts of human mental faculties are located in different regions of the brain. What Broca's case argued was that mental capacities are localized in separate regions of the brain; the locales that house specific abilities can be discovered, but the separation of human capacities physically realized in the brain is nothing like the traditional division of the mind into faculties, and nothing like Gall's divisions, either. With further discoveries, Broca's suggestion became a movement. In the 1870s Carl Wernicke described a patient with the obverse of Tan's disability. Wernicke's patient could not understand speech. At autopsy, the patient was found to have a brain lesion somewhat to the rear of Broca's area.

Meanwhile, similar developments were taking place in the study of patients with visual deficits. Various psychiatrists, including Wernicke and Theodor Meynert, one of Freud's teachers in Vienna, began to put these reports into a more or less systematic idea of how our brains produce our cognitive capacities. The near surface of the brain, the cortex, is a system of modules connected by nerve-fiber tracts. Each module does its piece, its process, its transformation, to the signals sent to it from other modules or from the senses, and subcortical nerve fibers pass on the result to other modules. The brain is an information-processing system with spatially localized cognitive parts that do special things to the information. (Karl Pearson compared the brain to a telephone exchange, but that was not quite the psychologists' model. Pearson, perhaps the worst philosopher there ever was, claimed that material things do not exist—there are only subjective experiences, produced by the *brain*.)

With the connectionist, modular picture, the job of psychology is to figure out where the modules are, what each of them does, how they are connected, and how their combined activity produces mental phenomena and actions. The way to do the job is just as Broca and Wernicke and others did it: see what normal abilities can be separated or "disassociated" by brain damage, so that a patient has one ability but not the other. Such a combination of capacities and incapacities in one person with brain damage means that the damaged region of the brain is critical to one of the abilities but not to the other. Autopsy will show the location of the critical module. "Double dissociations," in which one patient loses ability A but not B, and another loses ability B but not A, show that A and B have mechanisms that are, at some stages of processing, independent of one another. Thus, bit by bit, an architecture of cognition can be constructed. While he was a medical student, Freud learned the paradigm from Meynert, rejected the whole idea, and then adopted it in another form.

Freud was captivated by a somewhat different picture of how the brain produces the mind. Camillo Golgi had found that a silver chromate solution would color some nerve cells, and after examining them with the microscope he concluded that their connections formed a continuous web, which accorded with nineteenth-century ideas that the cortex is a unified system that works by continuously distributing or redistributing something throughout it. Using Golgi's stain, around 1889 Santiago Ramón y Cajal distinguished synapses and axons and showed that they do not quite meet. The brain looks like a big web, a network, in which the nodes reach out to one another but do not quite touch. Something, no one knew what, must pass between. Cajal took the structure as a key to how the brain works: there is no continuous fluid, harmoniously or inharmoniously distributed through the brain. Instead, the brain has parts, and the parts are nerve cells, and mental phenomena are produced by the parts somehow activating their neighbors. The influence of one cell on another is the very process of passing information.

Freud first bought into something like Golgi's picture, and in 1891 used it in a book, *On Aphasia,* to puncture holes in Meynert's project. *On Aphasia* is the best scientific work Freud ever did, and except for his essay on diseases of children, perhaps the least read. On thinking about Cajal's results, Freud's view changed to a combination of Cajal's information viewpoint and the more holistic viewpoint. Freud never published these later ideas, but he wrote them up as "Project for a Scientific Psychology" and sent the manuscript to his crank medical friend Wilhelm Fliess, whose wife later sold Freud's correspondence, which is why we have it still today.

In the 1890s Freud made still another turn. Denied academic advancement, he had made his living as a private-practice "neurologist"—we would say "psychiatrist." His first thought was that neurosis breaks the normal mind quite as much as brain damage does, and the structure of the mind can therefore be discovered from the behavior and testimony of neurotics, just as it can be discovered from the incapacities of aphasics. (Psychotics are *too* broken.) By 1900 he had begun to make famous the idea that the normal structures break down in other contexts as well, famously in the dreams and jokes of normals as well as neurotics—although Freud eventually seems to have thought that nearly everyone is mentally ill. Freud's goal was to identify a mechanism of mind without reference to the brain or to data on how the brain works and breaks.

Freud replaced brain regions with novel entities, the Ego and the Unconscious, eventually joined by the Id, the Unconscious Ego, the Pre-Conscious, and the Ego-Ideal, for all of which Freud denied a physical interpretation. In most of his psychoanalytic writing these entities played starring roles in an unphysical internal theater. Ultimately, he could not resist drawing them as spatial regions, and in his last works he assigned them places in the brain.

Freud's first psychoanalytic book, *The Interpretation of Dreams,* was not well received; from reviews it seems that many psychiatrists thought the book and the author had left science, which they had. The solipsistic logic of *The Interpretation of Dreams* and Freud's candid description of his methods in his published papers in the 1890s make it clear, as several scholarly studies have shown, that by 1899 Freud had segued from scientist to mountebank. He demanded that his patients come up with recollections that accorded with his hypotheses, told them what was required, and badgered them until they either produced what he wanted or left his services. In the mid-1890s Freud was naïve enough to describe his methods candidly; in later years he was sophisticated enough to hide that history.

What do you do when your ideas have a limited but devoted following, when, after having been given a scientific audience and tried out there, your claims and witnessing are no longer welcome in reputable scientific journals? You form your own "scholarly" society, have your own meetings, establish your own journals and outlets. That is just what Freud did, beginning with the formation in 1902 of what became (in 1908) the Vienna Psychoanalytic Society. Marginal scientific movements since have followed the model.

It cannot be said that Freud never looked back after 1900; on occasion he did, but he had either forgotten or deliberately dissembled about the descriptions of psychoanalytic method that he had published in the 1890s—intrusive, directive, dogmatic methods and poor results—and about what he had written in private. For all that, he attracted the adherence of some of the best and most critical minds of the twentieth century. Hans Reichenbach, perhaps the most empirically minded philosopher there ever was, credited Freud's theories and even lectured to psychoanalytic societies, although he never discussed Freud's work in print. Jared Diamond, who writes brilliant volumes on society and the environment and insightful essays on method, champions the effectiveness of psychoanalysis and ranks Freud with Darwin, which is something like ranking Erich von Dänikan with Isaac Newton.

The year 1904 also saw the beginning of psychometrics, an enterprise that ultimately has at least a methodological connection with Freud. The basic idea came from Charles Spearman, a British army officer who studied psychology in Germany and published his seminal paper while still a graduate student. By measuring the behavior of a lot of people in a variety of special tasks, we can gather statistics on the tasks. Performances on some of the tasks will be correlated: if the scores on tasks A and B are multiplied together for each person and then averaged over all persons, the result will be different from the result of multiplying the average score on A by the average score on B. A little manipulation of that difference yields the "correlation coefficient" relating tasks A and B for the population. So by having enough people do enough tasks, one can obtain an entire system of correlations, A with B, B with C, C with D, A with C, A with D, B with D, and so on. Correlations of tasks should have a causal explanation: A causes B, or B causes A, or some third thing, G, causes both. If A and B have a common cause, G, then, as the value of G varies from person to person, so should the values of A and B vary, with the result that A and B will be correlated. The principle is quite general: you can verify it anywhere you have two lights on the same switch circuit.

The tasks Spearman considered were those of early intelligence tests, which combined cognitive and physical tasks, for example memory tests with reaction-time tests. There is no reason to think that someone's performance on one of these tasks influences his performance on other tasks, or that one person's performances influence another person's performances, so if there is a correlation, there should be one or more common causes, which Spearman assumed to be features of the mind. But how many common causes? Spearman argued that we can identify a single common cause, which he called g, for "general intelligence," of all the task performances on an intelligence test. We can know this, he argued, because the correlations themselves satisfy equations involving four variables and four correlations—tetrad equations, he called them—that can hold only if there is a single common cause of the tasks.[1]

Subject to a lot of qualifications, Spearman was essentially correct *if* the equations he had in mind actually held in the data, which they did not, as his students soon found. They proposed ad hoc extra factors besides g to explain why not all of Spearman's predicted tetrad equations held in their samples. The proper scientific project would have been instead to characterize mathematically the alternative structures that could account for various possible sets of constraints like Spearman's tetrad equations, but neither he nor his students were up to that. Spearman's movement did not end well, either in methodol-

---

1. Let X_1, X_2, X_3, and X_4 be variable quantities, taking different values for different individuals. Let ρ_{ij} denote the correlation between X_i and X_j. A tetrad equation is any one of the form $\rho_{ij}\rho_{kl} = \rho_{ik}\rho_{jl}$, for example, $\rho_{12}\rho_{34} = \rho_{13}\rho_{24}$.

ogy or in substance. At Stanford, Truman Kelley, one of his most distinguished followers, concluded from intelligence tests with American children of Japanese descent—almost certainly all Issei or Nisei—that Japanese have inferior minds. Tell *that* to General Motors.

Aside from overlooking environmental explanations of test-score differences, Spearman and his followers had a further problem: there were way too many tetrad equations to test. With, say, an IQ test having a hundred questions there are 3,921,225 distinct possible tetrad equations. Nobody had enough female calculators for that. (Educated women were the scientific concubines who did the work in psychological laboratories until the digital computer displaced them in mid-century and women were allowed to become academic psychologists.) That problem was partially solved thirty years later with the advent of Leon Thurstone's "factor analysis." Thurstone's method was simpler, faster, and in a sense less perceptive than Spearman's. Spearman had identified a pattern of relationships among correlations that, wherever they might occur, indicated a single common cause of the correlated variables. Granted, he had done nothing to identify the patterns that might indicate the presence of more common causes or their relations, but for the singleton case he had done something. Thurstone's idea was to posit a single common cause and compute how much of the correlation among tasks could be accounted for on that hypothesis. Then postulate a second common cause and use it to explain as much of the leftover correlation as possible. Then postulate a third common cause and so on, until there is no more correlation left to explain. Thurstone had no proof that in general his method found the correct number of unobserved common causes, and there is no proof today, because it does not.

Thurstone's method could be done by hand more easily than Spearman's; it was general, and it applied no matter how many hidden common causes there might be. A leading psychology text soon recommended Thurstone's way on grounds of computational feasibility. The introduction of digital computers twenty years later made Thurstone's method a standard—now implemented in most commercial, computerized statistics software programs—that can easily and rapidly be applied to lots of variables.

Thurstone was disingenuous—or at least artful—in a way that had consequences. He titled his book *The Vectors of Mind*, as if it were a tract on the mathematical structure of the mind. His introduction said the book was merely about a technique for summarizing data. Psychometricians have exploited the equivocation ever since, implying on the one side that factor-analysis models find causes, while asserting on the other that they merely simplify descriptions of data. The result has been nearly a century of unproductive science. Methods for inferring causes need to be tested, mathematically or by simulation, for their accuracy at *inferring causes*. Statistical estimation techniques and data summaries need only weaker tests or none at all, because their ambitions are more modest. Various psychometricians devoted their careers to making claims that factor-analysis models explained behavior, without providing any analysis of how reliably and informatively the procedure uncovers causal mechanisms. Even after the introduction of the digital computer, which made simulation studies of the accuracy of factor analysis easy to do, almost none were produced. The evasion—or confusion—was epitomized by a psychometrician who pronounced that the latent variables found by factor analysis are not real, but that it is very important to use correct statistical methods to estimate them. Shades of Karl Pearson.

So the psychometric tradition reproduced the psychoanalytic tradition: equivocate about what is claimed and avoid stressful tests of accuracy and reliability. Throughout the twentieth century, psychometricians predicted essentially nothing about neuropsychological discoveries.

After the middle of the last century, cognitive psychologists developed the very kinds of theories—the spirit, not the letter—that Freud had proposed in the 1890s, and in many cases their methods had liabilities almost as serious. In 1990, without knowing of Freud's book, Martha Farah, a prominent neuropsychologist, revived Freud's arguments from *On Aphasia*. Farah's book about the causes of visual deficits uses the same forms of argument that Freud used about aphasia and defends essentially the same connectionist viewpoint. The two books provide a striking parallelism, so much that one would think Freud and Farah were in e-mail contact with a one-hundred-year delay. Just as *On Aphasia* was the best of Freud's scientific work, *Visual Agnosia* was among the best books of the late twentieth-century Freudian revival known as cognitive psychology.

Relatively cheap, if inconvenient, computers became available early in the 1960s, and they became increasingly cheaper, more powerful, and more convenient. A digital computer is a limited version of a universal Turing machine. So suppose you want a computer to simulate someone's behavior in a psychology experiment. You need only describe the behavior precisely in words, or in words and mathematics, and be a sufficiently clever computer programmer, and you can produce a simulation of the real behavior. The simulation may not do its simulated acts with the same speed as the real acts, but it will do them in the same sequence with the same simulated outcome. When psychologists realized as much, there was no stopping them from giving new life to Freud's ghost.

Some ambitious, insightful, and misguided psychoanalysts literally wrote computer programs to simulate the psychoanalytic processes Freud had postulated. Odd and uninfluential as such work may have been, it had a historical sensibility. In the 1890s, in his essay on aphasia and, especially, in his long unpublished essay "Project for a Scientific Psychology," Freud had developed a computational model of the brain/mind, but, of course, without any idea of a physical computer on which to simulate hypothetical states and processes. Academic psychologists followed a similar path.

Allen Newell was a big, smart, serious man who trained as a mathematician and went to work at Rand and then went into psychology. Newell realized that he could simulate laboratory behavior with computer programs. He further realized that a certain kind of programming system, called a production system, made the programming very easy. Production-system programs are simply a series of *If . . . then . . .* rules together with a bag of data. Whenever the *If . . .* part is recognized by the computer, the *then . . .* part is added to the bag, and the system goes on. Relatively little control structure is needed to make a production-system program; one doesn't need to think about subtle algorithms. So Newell developed a programming system called Soar, built on a production-system language, with special data bags for "short-term memory" and "long-term memory" and a very simple learning algorithm, with one twist: when certain kinds of events were discovered to occur together they could be collapsed into a "chunk," and associations among different chunks could be learned. Chunks that occurred together could be conjoined into super-chunks, and so on. With these pieces, Newell and his followers could simulate behavior in a lot of simple psychological experiments. They could predict almost nothing that was not already known independently or immediately deducible from the very fact that any finite description of a sequence of actions can be simulated in a computer program.

The Soar enterprise illustrated dysfunctions at the very heart of cognitive psychology in the period after 1960. Newell claimed he had a "unified theory of cognition," based on the fact that he did all of his simulations within a single programming language. But a programming language, even with a psychological gloss on some of its pieces, is not a theory, let alone a unified theory. An entire book of essays on the psychology of children's development claimed it was a unified theory, describing one experiment after another, simulating each result in Soar, and claiming the results as evidence that production systems are the programming system of the mind. Once upon a time I asked one of the editors to show me a result in the volume that could not be simulated in BASIC. He admitted, uncomfortably, that he could not do so. Newell formed his own Soar society, which for a while held regular meetings. As in Vienna, attendance and presentations were confined to believers.

Freud's early psychological model was the sort of thing that we now call a connectionist computer. John von Neumann, the mathematician who helped to invent both the modern digital computer and the theory of computation, wrote an elegant little book, *The Computer and the Brain*, arguing that brains are nothing like the kinds of computers we commonly use. Our computers are serial; they do one thing at a time. The brain, von Neumann argued, is a parallel computer that does a lot of things simultaneously and then somehow puts the results together. Connectionist computers are parallel computers: systems of nodes with on/off states, wired up with connections and programmed so that the state of any particular node—on or off—is determined by the states of the nodes with direct connections to it. Such computers are not common, but their behavior can be *simulated* by a serial computer. Under the name "neural net," such programs became another tool for psychologists.

Suppose you have a set of input nodes, each of which can take a value 0 or a value 1, and a finite set of output nodes with the same range of values, 0 or 1. Any function that determines the values of output nodes for every possible assignment of values to the input nodes can be computed by some neural net. To compute a particular function, the nodes have to be linked up correctly, often with intermediate nodes, and the strengths of influences of one node on another must be properly adjusted. There are computer programs that use sample data to automatically adjust these parameters. So if a psychologist wants to simulate how people learn from examples, she can do so with a neural-network program. If the psychologist wants to simulate how two concepts come to be associated, so that instances of one tend to co-occur with instances of the other, she can do so. Even if the psychologist wants to simulate the effects of brain damage, she can train a computer to do some normal activity—discriminate words from nonwords for example—then simulate brain damage by inactivating some of the nodes and obtain a simulation of some of the abnormal behavior of brain-damaged people.

The performance of neural-net psychological models can be astonishing and impressive. Besides that, they have a scientific credential: we think brains really do process information, enabling memory, deliberation, planning, wishing, and so forth, via signals among the nerve cells of the cortex—although sundry chemicals help. But there is a real scientific accomplishment, and not merely a technical performance, only if the specific mechanism of a psychologist's neural-net model corresponds to a specific mechanism common in people, or at least animals, and that correspondence is demonstrated. That connection is almost always missing in connectionist psychology. Studies of simple animals, such as the sea slug, produce a detailed wiring diagram showing how particular behaviors come about in response to external stimuli and showing how the animal learns. Studies of cells in the visual cortex show that neural spikes have a probability interpretation. Nothing like that has emerged from connectionist psychology. It has provided no striking predictions about human behavior and the mechanisms by which it is produced—only after-the-fact simulations of behavior, simulations that psychologists should have known could, with sufficient cleverness, always be produced. (A similar kind of now popular quasiscientific psychology treats human learning as a calculation of posterior probabilities by the Reverend Bayes's procedures, without regard to computational plausibility or biological realization.)

Under Newell's influence one of the most prominent contemporary psychologists, John Anderson, developed his own "cognitive architecture"—which is in some of its general features stunningly similar to Freud's "Project for a Scientific Psychology." The theory posits a network with activations passed from node to node and nodes that represent particular propositions or thoughts. The passing of activation is affected by the operation of production rules, and the network is organized into modules, such as "declarative memory." Anderson and his students estimated how long it took subjects in experiments to carry out simple steps in his system—the so-called "atomic components of thought." They implemented the results and the theory in a computer program, ACT*. But Anderson realized that his system, like Newell's and others', was radically underdetermined by the data. So he put it in suspension and went to work on theories postulating that, once hidden internal computational costs and environmental contexts are taken into account, human decision making is "rational" decision making of the kind Reverend Bayes proposed. But then, recognizing that this is not much of an empirical theory of human information processing, Anderson implemented improved learning procedures to form a new theory he called ACT-R. After the development of brain-imaging techniques that identified local regions of the brain involved in various particular cognitive tasks, he used conclusions from neuropsychology to map ACT-R modules to brain regions—almost a reenactment of Freud's attempt at localization at the end of his career, but with a difference. Anderson had previously estimated from psychological experiments the time required of each module in specific kinds of tasks (solving algebraic equations, for example), and he could therefore predict the time required for those brain regions to act in such tasks. Independently, while subjects were solving the equations, he estimated, from functional magnetic resonance images of their brains, the time delays in activities between the brain areas to which he had assigned his hypothetical modules. So Anderson has turned his theory into something testable and begun testing it. The results so far are mixed, but the effort is a step out of Freud's long shadow.

A historian, Stephen Stigler, once announced "Stigler's Law": nothing in the history of science is named for the person who discovered it. In 1895 Freud described a neural mechanism for learning. Nerve cells, he speculated, have "contact barriers" that inhibit the reception of signals from other cells. Once a signal from one cell breaks through the contact barrier at the synaptic connection with another cell, that connection is made easier in the future. Thus from experience, pathways of neural connections are formed and become more or less fixed. That is learning. I do not know whether Freud's proposal was original—and it was not published until 1950—but it anticipated by more than fifty years the work of Donald Hebb, to whom the idea is almost always attributed. More recently, Eric Kandel and his collaborators worked out the molecular basis of facilitation in the "Hebb synapse" and detailed the biochemical mechanism for simple learning in sea slugs.

We know how a sea slug learns to retract its hood in the presence of a stimulus, we know the wiring, we know the chemicals. Unless you want quantum mechanics, that is psychology from the very bottom up. Can we do the same for people? There are, for example, involuntary eye movements—saccades—that are almost perfectly predicted by the activities of clusters of brain cells in primates studied experimentally and so are presumably predictable in us. Fortunately or not, humans have a scope for learning and acting that is a lot wider than that of sea slugs and more complex than saccadic eye movements. Can we break up the many aspects of speaking or reading or counting or learning to speak or learning to count, let alone learning the skills of a novelist, so that they correspond, one to one, to combinations of identifiable physical features of neural processes, the way the sea slug's biochemical development corresponds to learning to retract its hood? That is the fundamental hope of psychologists and the deepest fear of philosophers.

12

What We Have Here Is a Failure to Communicate

People often do not say what they mean or mean what they say; sometimes they say too much, and sometimes they say too little. In science these lapses take a lot of forms, some of them funny, some of them a bit sad, some slightly infuriating. Examples follow.

Military Research

In 1962 the University of Montana effectively expelled me for refusing to show up for required Reserve Officer Training Corps courses. I could go to other classes if I wanted, but they would not count for graduation unless I did my soldier training. That's how it used to be in Montana. Some of my friends managed better. David Hunt, who was then a young Anglophile, showed up out of uniform for the weekly formation and parade—he wore a white suit and spats and carried an umbrella. With no brig available, the ROTC guys made a deal: David would not show up, and they would pass him. Jack Mueller, who had served in the Marines and was in the Marine Reserves but was required to take ROTC anyway, put a swastika armband over his army ROTC uniform and spent a day giving the *Sieg Heil* to everyone he came across on campus. As punishment, Jack's Marine Reserve commander, a Colonel Jack Ripper doppelgänger from Hamilton, south of Missoula (why is it always the south?), wanted to reactivate Jack. Jack got out of it when the officer asked him why he had gone on the one-man parade. "Because I'm a Marine, sir, and I couldn't stand wearing that stinking army uniform, sir." Jack stayed in college, skipped the rest of ROTC, and went on to write "June Is Dairy Month" ads for the Department of Agriculture.

So when I grew up I thought it was fair to take some of the military's research money, as long as I didn't have to make bullets or bombs or design improved tortures. My first acquaintance with military research came on a visit to Pensacola. I was talking with Ken Ford at the University of West Florida (you may snicker, but they had an undergraduate who came in second in college *Jeopardy*) when he received a call from a local defense contractor. Somebody in the contractor's outfit had accepted a contract for thirty grand to install software for a lieutenant at the Naval Aerospace Medical Research Laboratory (NAMRL is the charming acronym). The contractor figured the job would take about thirty minutes, and a thousand dollars a minute would just beg for a congressional investigation. He wanted Ken to clear the way with the base commander so the software could be installed for free. Ken had been a SEAL and had creds on the base.

So we met in a base conference room with the base commander, a redheaded captain, along with one of Ken's graduate students, the defense contractor guy, and two of his techs. The captain gave the techs and the graduate student directions to the lieutenant's lab, and off they went to install the software, for free. Ten minutes later they were back. "So fast?" we asked. No, the techs explained, the lieutenant said he had issued a *contract*, he wanted the software installed under the *contract*, and he would not let them install it for *free*. And you wonder about your taxes?

For a while I did contract work for the Navy Personnel Research and Development Center (NPRDC to you), then located on a lovely bluff above San Diego harbor. The navy wanted me to help find the answers to lots of questions about how personnel training and quality affected the performance of aircraft units, ships, air traffic controllers, and so on. It did not go well. One day, visiting the base, I met with the civilian in charge of my contract and a charming woman—an outside contractor—who had helped him win an award using some of my software. Obviously proud, he graciously, and probably truly, said she had done all the real work. We went to lunch, and the guy's boss happened by. After introductions, the boss remarked on the award. At some point I said, with a smile, "He says she did it all." I returned to Pittsburgh to find a message from the guy in charge of my contract: never set foot on the base again.

To figure out what made for better performance, I needed data on the performance of naval units. The navy runs war games with real ships, and the ships get scored on their performance. I had a clearance, but no one I could find in the navy would give me the data. One day I got a call from a woman in the Office of Naval Research who told me she could get me navy wargame evaluations if I would meet her in Washington. So I went to D.C. and met her for coffee. She was there with a guy she introduced as a retired navy commander. The guy could get me the data—for $20,000. He turned out to be her lover. I wish I had worn a wire.

The other things I needed to answer the navy's questions were the billets and competence evaluations for ships' complements. Then I could try to figure out how differences in personnel quality were associated with ships' performance—if I could get hold of measures of ship performance. The personnel data were classified, but despite academics' nervousness just about anybody can get a low-level clearance, even me, and I got the data, albeit in a sealed room. In those days every unit commander in the navy regularly evaluated every enlisted person in the unit on a 4-point scale, and I had the entire database for all enlisted personnel in the navy for several years. It did no good; 99.99 percent of all enlisted personnel were perfect 4s, always. The seven guys who weren't perfect 4s must have been colossal screwups or, more likely, typos. Nowadays the unit commanders have a limit on the number of 4s they can give, but that wouldn't have helped me much: I am pretty sure they always fill their limit.

Medical Research

An estimate in 1990 claimed there were 70,000 academic journals, give or take, and there are presumably a good many more than that as I write. It's hard to know how much they are read, but citation rates indicate that there is a pecking order: the *New England Journal of Medicine* is read a lot more than, say, *Artificial Intelligence in Medicine.* The density of journals means that in many areas, clinical medicine, for example, reviewers cannot feasibly determine whether a result has been published elsewhere or whether a contradictory result is in the literature somewhere. Nowadays there are computerized aids, but they are in many cases woefully insufficient, and even they take time. Peer reviewing is mostly volunteer work. The results of too many researchers, too much research money, too many journals, and pressed reviewers can be odd. Figuring out how to treat pneumonia is an example of odd.

Pneumonia takes a lot of lives and, before there was a vaccine, took more. When someone arrived at a hospital with pneumonia, physicians had to decide whether to hospitalize the patient or send her home with drugs. The effective difference was that if sent home the patient would take oral antibiotics, and if put in the hospital the patient would receive intravenous antibiotics. Intravenous antibiotics are more effective, but hospitalization is expensive, and a weakened hospitalized patient risks catching something else in the hospital. So the question is: how should physicians decide which patients to send home with a prescription and which to admit to the hospital?

Ideally, one would do an experiment matching a sample of people with pneumonia for other characteristics that might be relevant, such as age, and sending at random one member of each matched pair to the hospital and the other one home and seeing which group—hospitalized or at home—has the higher death rate. Not likely that could pass the ethics review board. (We remember the evil of the Tuskegee syphilis experiment, but we tend to forget that under modern ethics-review standards the Salk polio vaccine would have been delayed for years.) So the National Science Foundation and the Whittaker Foundation gave a bunch of us a grant to do a nonexperimental study to try to predict which patients would most benefit from hospitalization. The bunch was a group of biomedical computer scientists, a distinguished if old-fashioned computer scientist (née philosopher) who once upon a time practically invented applied artificial intelligence, a neural-net guy, a physician group from a medical school, and my group of philosophers pretending to be computer scientists. We would all separately have the data from pneumonia cases that arrived at the University of Pittsburgh hospitals. We would know the facts about them on initial medical records, the follow-up records, whether they were sent home or admitted, and whether they lived or died in the course of treatment. The main job was to discover a procedure to predict the risk that an admitted patient would die.

It was a lot of data in a variety of formats, and since the problem was serious we all worked very hard on it. The old-guy computer scientist produced what is known as a "rule-based system": if a patient has this feature and that feature but not that other feature, then the patient will die. The rule-based system was a really terrible predictor, but the computer scientist, who had put it together by asking physicians what rules they would use, reported that those he consulted really enjoyed trying to think of rules. Nothing like changing the point.

The physicians used a tried and sometimes true statistical method, logistic regression, a form of regression adapted for predicting variables that have only two values, like life and death. Their predictor was better than the rule-based predictor, but next to last in accuracy among all the methods produced by the five groups. Bayes nets are a kind of compromise medium between neural nets and logistic regression. The variables are linked together, in some ways like the nodes in a neural network, but the structure of the linkages constrains the possible probability relations among those variables. By searching the data for such constraints, one can discover the structure of Bayes networks that can explain the data. Bayes nets cannot be discovered for every problem, but they are more flexible than logistic regression. The biomedical group developed a Bayes net predictor that predicted better than the logistic regression predictor. Using a different way of building a Bayes net, my group developed a predictor that did still better. The neural-net group did best of all.

So the order of accuracy from worst to best was: rule-based, logistic regression, Bayes net 1, Bayes net 2, neural network. We published the results together in a long essay in *Artificial Intelligence and Medicine*. Soon after, another essay appeared in the *New England Journal of Medicine*, reporting a pneumonia triage procedure based on the physicians' logistic regression method. How that fared against alternative predictors was not mentioned, nor were the alternatives themselves.

Wildfire

Wildfires occur year-round in the United States, they do a lot of damage, and the natural impulse is to try to put them out as soon as possible. Firefighters would like to have at least a week's warning—two weeks or a month would be better—as to where a fire is likely to break out so that equipment and people can be positioned not too far away.

Wildfire is produced by some ignition source striking dry deadfall. The amount of deadfall, how dry it is, the temperature, the humidity, the kind of live vegetation and how dense it is, and the winds largely determine how big the fire will get if neither rain nor humans intervene. Ignition is a not-quite-random event—thunderstorms can be inaccurately predicted not many days ahead, but campers and arsonists and the occasional fool running on his rims and sending sparks into the brush (yes, really happened in Idaho) cannot be. Prediction accuracy has dimensions of space, intensity, and time: the smaller the interval of time, the smaller the area, the more exact the fire acreage, and the more in advance of the event, the more unreliable any prediction will be. What is wanted is the most accurate prediction of burn area for the nation divided into the smallest regions, for each day, as far in advance as possible. Such multiple desires cannot all be maximized, and so there are trade-offs.

As with climate, there are two approaches to predicting wildfire, the physical and the statistical. Physical methods attempt to obtain explicit equations relating fire occurrence or size or speed of spread or something to physical characteristics of ground cover—moisture, species, density, and so on. Statistical methods look at historical data on where and when wildfires occur, the fuel type (woods, grasslands, alpine forests), temperature, moisture content, perhaps previous weather conditions, and then use a statistical or machine-learning algorithm to attempt to forecast wildfire.

However it is done, the essential thing is to *test* the method of prediction to get an idea of the reliability of the forecasts, the trade-offs involved, and whether the effort that produced the forecasting method was worth the bother. There are two standard ways to indicate how well a forecast is doing. If the predictions are simply whether, for various areas and times, a fire (or a fire greater than a given size) will or will not occur, then the standard is the "confusion matrix," a fancy word for a table that has four entries: how many times a fire was predicted and a fire occurred, how many times a fire was predicted and did not occur, how many times a fire was not predicted and did not occur, and how many times a fire was not predicted and did occur. Like weather predictions, some fire forecasting methods assign probabilities to the occurrence of fires. The errors in forecasts have two directions: fires that occur but were not predicted, and fires that were predicted but do not occur. In using such probability forecasts to make decisions (such as where to deploy the firefighting equipment), the errors on one side will tend to decrease as the forecast level of probability used to decide there will be a fire increases, but the errors on the other side will increase. If the fire managers decide to deploy equipment only when the forecast gives a high probability, then when the equipment is deployed there will often be a fire, but the managers will fail to deploy equipment in many cases (those with lower forecast fire probabilities) where fires do actually occur. Work on radar detection in World War II produced a way of showing this trade-off, the "receiver operating characteristic," or ROC, curve. The ROC curve is just a plot of the forecast probability on two axes, one the percentage of correct predictions of fire and the other the percentage of incorrect predictions of fire. Different points on the ROC curve represent different trade-offs between true positives and false positives. An entire confusion matrix can be calculated for any point on a ROC curve. The area under a ROC curve, or AUC, gives an overall estimate of prediction accuracy. If prediction were perfect, that area would be 1. If prediction were just random, the ROC curve would be a straight line at a 45-degree angle and the AUC would be $\frac{1}{2}$. For any probabilities below that straight line, one would do better using the *opposite* of the prediction.

Some forecast assessments use a "calibration curve." The idea is that when a weather forecaster says "The probability of rain is X" and the percentage of times it rains is 100X, then the weather forecaster is "well calibrated." But calibration alone doesn't measure the accuracy or usefulness of a forecaster. Ideally we would like a forecaster who says only that the probability of rain is 1 or that it is 0 and is right every time.

Providing a ROC curve or confusion matrix is just one part of assessing a forecasting method. The other part is assessing whether there is a better method or whether utterly simple methods are just as good. I once was a member of a committee to examine a doctoral thesis that aimed to determine whether proteins, which can be very complicated molecules, belonged to a common chemical family. The candidate had developed an elaborate statistical procedure, but it came out in the examination that an old, simple matching algorithm, known as BLAST, did better. There's a lot of that going around, and it applies to wildfire prediction as well. For example, at any season of the year, wildfires generally occur close to where they occurred in years past (say within five to ten linear miles of the edge of a previous fire). For wildfire, it's location, location, location. So an important question to ask any wildfire forecasting method is whether it does better than just predicting fire based on what happened nearby at about the same time in years past. Still other issues are the data and the variables. There is an elaborate system of ground-based measurements of precipitation, temperature, humidity, and other variables, and that is one source of data. Satellites measuring the spectrum of radiation from the surface of Earth are another source. Teams of experts have turned these satellite measurements into "products" that try to measure attributes of the ground vegetation cover as well as temperature and other variables. Fire occurrence histories are available from Forest Service station reports and can also be estimated from satellite data. Whether surface data or satellite data gives the better predictions is an interesting issue. My experience, to NASA's disappointment, is that ground-based data are more informative than satellite data, but a combination works best.

I worked on forest fire prediction for a couple of years, but the work was never satisfactorily finished. (My programmer left his wife, went flaky, and went off to the Burning Man.) The Forest Service has divided the United States into various fuel regions, and developing separate prediction methods for each such region helps prediction accuracy. For some of these fuel regions, reasonable predictions are possible a month in advance of a four-week interval for areas as small as twenty-five square miles. Any number of methods work about equally well. Our forecasts used a lot of ground-based data and some satellite variables, but fires can be predicted, although not quite as well, using only the history of nearby burns.

The federal government has a host of agencies interested in gathering data on wildfire, fighting fires, predicting fires: the Bureau of Land Management, the Forest Service, the Fish and Wildlife Service, the Bureau of Indian Affairs, the National Interagency Fire Center in Boise, Idaho, and doubtless more. Occasionally, the National Oceanographic and Atmospheric Administration and even NASA get into the business, since they launch or manage the nonmilitary satellites that collect data around the entire world every day. Each of these agencies has its subagencies and research projects for forest fire prediction. There is no such thing as a national competition to find the best fire prediction method, so each group continues on, some productively, some down a hole.

Sometimes the forecasters get together and report their methods and forecasts. At a meeting in Boulder a couple of years ago, I listened in. A few years before, I had asked a Forest Service group from Corvallis, Oregon, that regularly sent out fire forecasts whether they had statistical tests of their method. The reply was no, but they were working on it. In Boulder, their representative presented their physical model of fire, but still no real statistical tests of its forecasting accuracy. They were working on it. A group from Montana and the Dakotas was a good bit better. Bob Burgan is retired from the Forest Service in Montana, and he looks it—from Montana, not retired. He collaborated with two other retired Forest Service workers and others to produce and test a forecasting algorithm using satellite data. Their approach was statistical and quite sensible, but their predictions gave short notice—a week ahead—and only for huge areas of the country. Their calibration for predictions of at least one large (greater than 5,000 acres) fire for each of these vast regions within a week was very good, but no ROC curve or confusion matrix was provided; and they made no comparison with very simple forecasts, for example using just the history of burns and their locations.

Later in the day, as if Burgan had never spoken, a guy from the Bureau of Land Management gave a talk in which he said nothing much had been done about statistical fire forecasting. Then he introduced a group from a commercial statistical outfit, SAS. SAS takes academic statistics and turns them into a push-button computer program. The SAS guys said they had never done any work on forest fire prediction, but they explained that if the audience would find the money to pay them, they would. Burgan was bemused.

The Forest Service used to operate fire-watch towers in the western forests. Most of them are still standing, and the Forest Service will rent them out in the summer. Sick of computers and statistics, I rented a tower outside of Missoula for a week. The tower still had most of its equipment, chiefly a circular map of the mountains around it, so that a spotter could identify the location of a fire. The first night, a Saturday, I spotted a fire. I called the Forest Service. No answer. I tried the Bureau of Land Management. Closed for the weekend. I tried the Missoula fire department—not *their* department. Eventually I got hold of a county sheriff, and the next morning helicopters were dropping flame retardants. Communication is hard.

Light in Shallow Waters

Just about the most closely measured strip of land on Earth is around the mean high-tide mark on the American coast, because that's where the property values start. The property values aren't worth much if storms destroy beaches, because then they will soon destroy whatever is just beyond the beach. One way to reduce the effect of storms is to replace the sand they move, onshore and just off, in the littoral. With typical American efficiency, "beach nourishment," as it is called, is the responsibility of an entire fleet of government agencies: the U.S. Army Corps of Engineers (USACE), the Federal Emergency Management Agency (FEMA), the National Oceanic and Atmospheric Administration (NOAA), the U.S. Geological Survey (USGS), and the Minerals Management Service (MMS). Moving sand is the job of the navy. To move sand, the navy has to know where to get it.

So it happened that shortly after Hurricane Katrina, I and two of my colleagues from Pensacola met with a large group at Kessler Air Force Base, near Biloxi, Mississippi, to work on a contract to help the navy determine where the sand is, which amounts to estimating the depth of the seabed offshore and the mineral composition of the seabed surface. The measurement technology is LIDAR (Light Detection and Ranging). The idea is that two radiation signals of different frequencies are sent from above the water, usually from a small airplane, toward the surface. One kind of radiation bounces off the water, the other penetrates the water and bounces off the seabed or other object beneath the surface. The difference in time of return of the signals can be used to calculate the depth of the water above the seabed or above an object beneath the surface. LIDAR can discriminate quite small objects and terrain changes, and just as with other reflectance spectrometry, it can be used to determine mineral composition. The job was to turn LIDAR measurements into sand maps.

The meeting was led by a navy guy, who wanted sand. The discussion was mostly the usual silly stuff about what computer formats to use or not use, but one fellow insisted on a point: why just measure sand? While we were at it, we could develop software to detect obstructions and identify them, even to detect mines. We could map submerged seawalls. The navy guy said he would think about it and get back to us (he did not), but the focus was sand.

At Kessler, my colleagues and I met separately with Jan Depner, who had developed software for looking at LIDAR depth images mapped out as so many points on a computer screen. Depner is a guitar-playing guy who goes around the world LIDARing coastlines from a small airplane, and writes code in between. Ace job. Depner had a problem. Some LIDAR points were anomalously shallow compared to their neighboring points. These were "targets"—potential obstructions. The problem was that the images had to be looked at visually, and each pixel considered and debated: was it an obstruction or not? That took time and was tedious, and Depner didn't seem like a guy who liked tedium.

Back in Pensacola, we decided to solve Depner's problem. The three of us worked out an algorithm. Implementing it fell to the programmer who previously had gone off to Burning Man (I'm a slow learner), and sure enough, two weeks before the code was due, he went to see his mother, or something, leaving a mess. My other collaborator, Choh Man Teng, was (and is) a kind of genius. She can do just about anything with a computer but much prefers to paddle a kayak. Getting her down to work is hard, and she won't do trivial projects, but this one mattered and she took over, reprogramming (and improving) the algorithm from scratch. Tested on two different large regions her code had not previously seen, her program found all of the targets that Depner had independently identified by visual inspection, and no false positives. She added a program to outline the obstructions.

We sent the program and description to the navy guy and heard nothing back. He wanted sand; maybe he pounds it. Teng's code went somewhere, but probably nowhere useful. Where does software go to die?

13

Where Software Goes to Die

Michael Faraday discovered just about everything fundamental about electricity and magnetism except the equations. His nineteenth-century audience could verify most of his claims directly themselves or watch him demonstrate them in his public lectures. Later in the century Heinrich Hertz verified James Clerk Maxwell's prediction of electromagnetic radiation by moving a spark gap around a room. Had you been there, you could have watched the demonstration. Early in the twentieth century, a young American physicist from Baltimore, Robert Wood, refuted the claim of an eminent French physicist, René Blondlot, to have discovered a new form of radiation emanating from living matter and refracted by metallic prisms. Wood went to Blondlot's laboratory and during a demonstration, without Blondlot's knowledge, removed the aluminum prism that was supposedly bending the supposed N-rays, as Blondlot had named them. Reporting the spectrum, the French physicist gave the same numbers as he had when the apparatus was intact.

None of this works with climate change or the conclusions of large accelerator experiments or even evolution. The causes of climate change cannot be demonstrated in a public lecture. Pictures of glaciers can be shown, computer simulations can be run, conclusions and warnings can be shouted, but the thing cannot be demonstrated in an auditorium. No one in the public can check the calculations that go into climate simulation programs. An outsider cannot, for example, get the code for the Hadley climate simulations. An audience can see natural selection at work—watch a bird catch butterflies of one color but not another—but it cannot see evolution at work.

Science, like all human relationships, has always depended on trust. But when claims could be demonstrated before the eye, when experiments could be repeated with a few materials, what was trusted was not too hard to verify. That has changed, and now trust is much more essential, much more diffused, and more difficult to warrant. Requiring more trust produces more distrust.

Increasingly, in cellular biology, climate science, and other sciences, inferences depend on computer software. Often that software is tailored to a particular problem, and without it the inferences cannot be independently checked. Hell is other people's code, which means in practice that for software to be checked by outsiders requires that it be both public and documented. That too rarely happens except with commercial programs, and research grants seldom make available funds for documenting code.

For the scientific community, the problems of trust and repeatability have solutions, or partial solutions. For the lay community, not so much. Cellular biologists have done a remarkable job of organizing and making data available to all, and a version of their practice could usefully be adopted by social scientists and epidemiologists. The National Science Foundation and the National Institutes of Health could require that software used in publications be documented and posted and could regularly provide funds to do so. Gaining the trust of those outside the professions is more difficult.

Government must inevitably step into science when public health is at stake, and doing so puts the power of the police behind scientific conclusions. Vaccinations are required by law in schools. Prayer is not recognized as an effective substitute for scientific medicine, and where the health of children is at stake, refusal of scientifically endorsed medical treatment in favor of prayer is a crime. On scientific grounds, governments recognize smoking as a threat to health and restrict it.

Even without legal sanctions, the endorsement of governments and international quasigovernments like the United Nations carries extrascientific weight. Those endorsements may be founded on extrascientific reasons or on the views of one or a few politically powerful persons. Sometimes those views and biases are correct. I am convinced that the famous expeditions of 1919 to test the general theory of relativity were in part motivated by the desire of a few internationalist and pacifist British astronomers to vindicate Einstein in the midst of anti-German sentiment at the close of World War I. I think—although some competent scholars disagree—that their analysis of the data was biased by that goal, but as subsequent observations and experiments showed, Einstein was right about the path of light in a gravitational field. In 1965, on what was in retrospect insufficient evidence, the surgeon general endorsed the proposition that smoking has damaging health effects. Subsequent evidence has overwhelmingly confirmed that judgment. By contrast, the South African government's pronouncements about the causes and treatments of AIDS infection are an appalling example of governmental scientific opinion. Bert's complaint that the Intergovernmental Panel on Climate Change willfully ignored the role of a major factor in global warming, human population growth, seems correct. The panel produced several alternative emission estimates based on various projections of population growth but pointedly made no point of the matter. The panel's volume on mitigation acknowledges that population growth and GDP per capita were the main drivers of global carbon dioxide emissions in the twentieth century but provides not a word on policies to mitigate population growth. Jeffrey Sachs, our foremost academic campaigner against world poverty, initially ignored population issues, but to his credit he then produced an aggressive, humane, and feasible proposal for reducing population growth. Would that the scientists summoned by the United Nations to report on climate change had been so brave and so candid. I have no solution to collective scientific censorship except yelling.

The conjunction of statistical and probabilistic reasoning is too difficult for people, even people who claim to be experts. The sad story of

Herbert Needleman's trial illustrates some of the confusions and contradictions that abound in floating standards of "best practice" in statistical analyses for causal conclusions, a subject that merits a boring book all by itself. Causal analysis in many social sciences is effectively computerized rituals with factor analysis and regression. Better computational procedures are available for the problems, but bad methods persist by tradition and because bad methods give definite answers where better-founded procedures would report uncertainty. Uncertainty is hard to publish. But no computerized methods warrant failing to think about alternative explanations that the computer could not have found or failing to check them out where one can.

I see no way to make thoughtfulness a policy, but some of our educational practices tend to discourage it. Introductory statistics is taught in essentially all colleges, and in some is required of undergraduates. It is almost always separated from any careful, extended discussion of causality, even though the typical statistical applications are in aid of attempting to find what causes what. I have not touched on education curricula, but some things are absurd: in universities we teach students logic, although outside of mathematics they almost never meet a purported deductive inference. What they do meet, everywhere, all the time, are inferences about causes.

Scientific fraud is policed and therefore is risky business, but other forms of scientific misdirection are rarely recognized. There is nothing to be done about that except, where noticed, to take exception, preferably publicly. The failure to test claims for truth is better noticed than the failure to test methods for reliability.

Something about software that searches for scientific explanations invites neglect—the software makes no specific scientific claim, and it cannot speak for itself. It needs a spokesman and a minder. I am reminded of a story about one of my colleagues, Peter Spirtes. Years ago Spirtes worked for Carnegie Mellon's artificial-intelligence laboratory developing a program (called GREASE) for an oil company to help its salesmen select appropriate cutting oils for various kinds of machinery. The oil company was subsequently bought by Chevron. Some years later I gave a talk at the Chevron Research and Development

Center in California. At lunch one of my hosts asked if I knew Peter, and I said I did. The conversation went like this:

Host: He developed the GREASE program for the company we bought. I used to work there.
Me: Did GREASE work at all?
Host: Oh, yeah. It made my old company millions of dollars.
Me: And Chevron got the program when it bought the oil company?
Host: Yep.
Me: So Chevron sells cutting oils. How does it work for you now?
Host: It doesn't. It's on a shelf somewhere. Nobody at Chevron ever tried it.

So now you know where software goes to die.

Suggested Reading

WHAT THIS BOOK IS ABOUT

Another view of what is wrong with Father Callahan's disproof of non-Euclidean geometry and a précis of the book are available in Underwood Dudley's survey of mathematical cranks.

Callahan, J. D. *Euclid or Einstein.* Devon-Adair, 1931.
Dudley, Underwood. *Mathematical Cranks.* Cambridge University Press, 1992.

2. DAVID KORESH MIDDLE SCHOOL

Professor Darling-Hammond began criticizing Teach for America soon after it started operating, and she has kept up the attack. The Hoover Institution response is one of many.

Darling-Hammond, Linda. "Who Will Speak for the Children? How Teach for America Hurts Urban Schools and Students." *Phi Delta Kappan* (1994): 21–34.
Darling-Hammond, Linda, Deborah Holtzman, Su Jin Gatlin, and Julian Heilig. "Does Teacher Preparation Matter? Evidence about Teacher certification, Teach for America and Teacher Effectiveness." *Education Policy Analysis Archives* 13 (2005): 1–48.

Raymond, Margaret, Stephen Fletcher, and Javier Luque. "An Evaluation of
    Teacher Differences and Student Outcomes in Houston, Texas."
    CREDO, Hoover Institution, Stanford University (2001): http://
    credo.stanford.edu/downloads/tfa.pdf.

3. The Computer in the Classroom

My account of John Anderson's experience with computerized mathematics
instruction in public schools is based on a lecture he gave at Carnegie Mellon
University on the occasion of Herbert Simon's eightieth birthday. An indication
of his perspective can be found in the following:

Anderson, John, Lynne Rehder, and Herbert Simon. "Situated Learning
    and Education." *Educational Researcher* 25 (1996): 5–11.
Sandberg, Jacobijn, and Yvonne Bernard. "Interviews on AI and Education:
    John Anderson and Clotilde Pontecorvo." *AI Communications* 5
    (1992): 28–34.

4. Cosmic Censorship

Robert Pennock's book provides a careful assessment of intelligent-design claims.

Pennock, Robert. *Tower of Babel: The Evidence against the New
    Creationism.* MIT Press, 2000.

For a number of examples illustrating the unification that scientific explanations
provide—and God explanations do not—see the following:

Glymour, Clark. *Theory and Evidence.* Princeton University Press, 1980.

5. The Greatest Chemical Engineer There Ever Was

Useful biographies of the principals involved in the story of tetraethyl lead
include these books:

Boyd, Thomas. *Charles F. Kettering: A Biography.* Beard Books, 2002.
Hamilton, Alice. *Exploring the Dangerous Trades: The Autobiography of Al-
    ice Hamilton.* M. D. Miller, 2008.
Kimes, Beverly. *Engineers and Scoundrels: A History of the Automobile.* SAE
    International, 2004.
Midgley, Thomas. *From the Periodic Table to Production: The Life of Thomas*

Midgley, Jr., the Inventor of Ethyl Gasoline and Freon Refrigerants.
Stargazer, 2001.

Sicherman, Barbara. *Alice Hamilton: A Life in Letters.* University of Illinois
Press, 2003.

George Otis Smith's essay can be found in the *National Geographic* archives.

Smith, George Otis. "Where the World Gets Its Oil: But Where Will Our
Children Get It When American Wells Cease to Flow?" *National Geographic,* February 1920.

6. Galileo in Pittsburgh

The original report of the hearing board in the Herbert Needleman case is not
readily available. The account of Clair Patterson's career and travails by one of my
colleagues is both touching and shocking.

Davidson, Cliff. *Clean Hands: Clair Patterson's Crusade against Environmental Lead Contamination.* Nova Science, 1998.

The theoretical basis of Scheines's program, the difficulties with regression, and
the details of Scheines's analysis of Needleman's data, as well as further references
can be found in the following:

Fienberg, Stephen, Clark Glymour, and Richard Scheines. "Expert Statistical Testimony and Epidemiological Evidence: The Toxic Effects of
Lead Exposure on Children." *Journal of Econometrics* 113 (2002):
33–48.

Spirtes, Peter, Clark Glymour, and Richard Scheines. *Causation, Prediction,
and Search.* MIT Press, 2000.

Needleman has written a number of essays about the affair. Statements by both
Needleman and Ernhart can be found in the following article:

Ernhart, Claire, and Herbert Needleman. "Lead Levels and Child Development." *Journal of Learning Disabilities* 20 (1987): 262–65.

7. Bert's Buick

The scientific literature on climate and climate change is too large for any one
person to survey. With exceptions, we depend on review literature. The best
review of the state of climate science as of 2007, the effects of climate change, and
the prospects for mitigation is the three-volume report of the Intergovernmental
Panel on Climate Change. The panel, a continuing body sponsored by the UN,

plans further reports in 2010 and 2014. Jeffrey Sachs considers poverty, population, and climate in the following book:

Sachs, Jeffrey. *Common Wealth: Economics for a Crowded Planet.* Penguin, 2009.

8. SACRIFICE OF THE LAWN

Discussions of vegetarianism, animal rights, ethanol production, and so on are so common that readers will need no guide to them, but one forgotten classic deserves remembering and reading:

Veblen, Thorstein, *The Theory of the Leisure Class.* Macmillan, 1899.

9. JOHN HENRY AT NASA

I find no scientific or economic issue that would warrant the cost of manned exploration and occupation of the moon rather than robotic exploration or none at all, but Harrison Schmidt has made the contrary argument.

Schmidt, Harrison. *Return to the Moon: Exploration, Enterprise, and Energy in the Human Settlement of Space.* Springer, 2006.

10. TOTAL INFORMATION AWARENESS

For details of the Ali Baba set analysis and for various discussions of how computer science is used and abused in intelligence work, see the following:

Kott, Alexander, ed. *Information Warfare and Organizational Decision Making.* Artech House, 2006.

11. IN AND OUT OF FREUD'S SHADOW

Anderson, John. *The Architecture of Cognition.* Laurence Erlbaum, 1995.
Farah, Martha. *Visual Agnosia.* MIT Press, 2004.
Freud, Sigmund. *The Standard Edition of the Complete Psychological Works of Sigmund Freud.* Edited by James Strachey. Hogarth, various years.
Freud, Sigmund. *On Aphasia.* Imago, 1953.
Hebb, Donald. *The Organization of Behavior.* John Wiley and Sons, 1949.
Kandel, Eric. *In Search of Memory: The Emergence of a New Science of Mind.* W. W. Norton, 2007.
Kelley, Truman. *Crossroads in the Mind of Man.* Kessinger, 2007.

Klahr, David, Pat Langley, and Robert Nechles, eds. *Production System Models of Learning and Development*. MIT Press, 1987.

Newell, Alan. *Unified Theories of Cognition*. Harvard University Press, 1994.

Pearson, Karl. *The Grammar of Science*. Cosimo Classics, 2007.

Spearman, Charles. *The Abilities of Man*. Pierides, 2008.

12. WHAT WE HAVE HERE IS A FAILURE TO COMMUNICATE

Cooper, Gregory, et al. "An Evaluation of Machine-Learning Methods for Predicting Pneumonia Mortality." *Artificial Intelligence in Medicine* 9 (1997): 107–38.

Fine, Michael, et al. "A Prediction Rule to Identify Low-Risk Patients with Community Acquired Pneumonia." *New England Journal of Medicine* 336 (1997): 243–50.

Index

Lavoisier, Antoine, 2
lead: concentration, 50–51; effect on
    children's intelligence, 51–52, 62, 63
LIDAR, 132
Limbaugh, Rush, 17
Luque, Javier, 142

machine learning, 106
malnourishment, 83
Maxwell, James Clerk, 134
mechanism of mind, 112
medical research, 125–127
Meynert, Theodor, 110
Midgley, Thomas, Jr., 43–48
midnight basketball, 17
military research, 122–127
mineral identification, 94, 96–99
Montreal Protocol, 47
moon, reasons for going to, 100, 144
moral responsibility, 84–85
Mueller, Jack, 122–123
multiple comparisons, 58–60

Nagin, Ray, 18
National Security Agency, 103–104
Nechles, Robert, 145
Needleman, Herbert, 51–52, 53, 55–57,
    60–63, 143
neural net, 119
New York Times, 81, 103, 107
Newell, Allan, 117, 145
Newton, Isaac, 2, 35–36, 69, 96, 113
North, Oliver, 102
Nozick, Robert, 87

Office of Scientific Integrity, 57
oil refining, 42
oil spills, 41
Owen, Gwil, 7
ozone, 47

Pathfinder, 91, 97, 99
Patterson, Clair, 50, 51, 143
Pearson, Karl, 110, 116
Pennock, Robert, 142
personnel training, 123–125
pigs, 8, 86–87, 88
planetary science, 3, 94–99
pneumonia, 125
Poindexter, John, 102
Pontecorvo, Clotilde, 142
Popper, Karl, 36
population growth, 71–76, 85, 87,
    136
principals (school), 18, 23, 24
production system, 117
programming languages, 95
psychology, 18, 32, 108–121
psychometrics, 113–116

Ramon y Cahal, Santiago, 111
Ramsey, Joseph, 97
Rawls, John, 9
Raymond, Margaret, 142
Reagan, Ronald, 65
refrigerators, 45–47
regression, logistic, 126–127
regression, statistical, 2–3, 21, 52–58, 61,
    63
Rehder, Lynne, 142
Reichenbach, Hans, 53, 113
Remote Agent, 92–93
ROC curve, 128
Rockefeller, John D., 48
Rorschach test, 108–109
ROTC, 122–123
Rouquette, Nicholas, 92n
Roush, Ted, 95, 98, 99

Sachs, Jeffrey, 74, 75, 136, 144
Salk vaccine, 125